SWITCH CRAFT

SWITCH CRAFT

THE HIDDEN POWER
OF MENTAL AGILITY

Elaine Fox

HarperOne
An Imprint of HarperCollins*Publishers*

HarperCollins books may be purchased for educational, business,
or sales promotional use. For information, please email the
Special Markets Department at SPsales@harpercollins.com.

Originally published as *Switchcraft* in the United Kingdom
in 2022 by Hodder & Stoughton

First HarperOne edition published in 2022

FIRST EDITION

Library of Congress Cataloging-in-Publication Data has been applied for.

ISBN 978-0-06-303008-4

22 23 24 25 26 LSC 10 9 8 7 6 5 4 3 2 1

CONTENTS

THE SECOND PILLAR OF SWITCH CRAFT
Self-Awareness

THE THIRD PILLAR OF SWITCH CRAFT
Emotional Awareness

THE FOURTH PILLAR OF SWITCH CRAFT
Situational Awareness

"Clinging to the past is the problem. Embracing change is the answer."

Gloria Steinem

"Freedom and happiness are found in the flexibility and ease with which we move through change."

Gautama Buddha

INTRODUCTION

I lay on my bed sobbing uncontrollably.

Crying was rare for me, but I was overcome by the enormity of the mistake I had made. I was seventeen years old, and several weeks before, I had decided not to apply for a university place. Instead, my plan was to train as an accountant so that I could make enough money to travel the world. But after several weeks as a work-experience trainee in a local accounting firm, I knew I had made the wrong decision. It felt like I had destroyed my future.

Everyone was perfectly friendly but I found the office staid and the work soul-destroying. Every day I would stare out the window, counting down the minutes until 5pm so that I could leave. I knew I could not stick with accounting but, coming from a low-income, albeit very supportive, working-class family in Dublin, my options felt very limited. Education seemed like my only escape, but I had realized too late, and the application deadline for university was tomorrow at noon. The Central Applications Office, which handled applications for every university in the country, was in Galway, on the other side of the country, and the final date to mail the forms had long passed.

I buried my head back into the pillow, until a gentle knock on the door jolted me out of my misery. My parents had never seen me so upset, as I told them how I had missed my opportunity.

"Well, actually, you haven't," my mother said.

I was stunned when she suggested we get a late train to Galway, stay overnight, and deliver the forms by hand the next morning. This positive thinking was completely out of character for my mum, who typically zoned in on problems rather than finding solutions, but my genuine distress seemed to push her into action. Before I knew it my dad had driven us to Heuston station on the other side of Dublin, and I was sitting on the Galway train filling out my application forms. My mother and I stayed in a tiny B&B that night and we had fish and chips in a busy restaurant overlooking the sea that evening. I still vividly remember the joy I felt the following morning when we found the applications office and handed over my sealed envelope.

Six months later, following much intense study to make sure I got the grades, a letter arrived to offer me a place in general science at University College Dublin. And that began an incredible journey into academia that I am still exploring. When I look back over those forty years since leaving school, it's incredible to reflect on the many twists and turns my life has taken along the way. I have had many highs and also many lows. Every transition has required multiple adjustments that forced me to change as a person—both internally and externally—in order to cope and adapt. For instance, as a shy teenager I would rarely take center stage and was terrified about speaking to groups of people. I had to work hard to overcome this fear of public speaking to become a university teacher, science communicator and life coach to numerous elite athletes and businesspeople to help them reach the top of their game. There is little doubt that my interest in studying the psychology of adaptability and resilience, which has become a lifelong passion, was forged by these early experiences. And of course, I now realize that even if I had missed that deadline for university, there would have been a way around it, or I could have taken another path altogether. Life is often about opening up to new possibilities and being able to see routes around obstacles and setbacks.

Navigating your future

There are always choices to make in life, and they are rarely "right" or "wrong." Whatever the situation you might find yourself in, there will almost certainly be many options that are hard to choose between. This natural uncertainty is a fact of life. Even when you look back, you can never be truly certain that you made the right decision. You might be grateful that you married the person you did, for instance, because you have such great kids and a happy life. But if you had married that other girl or guy, you may have had equally great kids and may have been even happier. You will never know. And this can be liberating.

Whether it is to do with career paths or personal decisions, there are lots of pathways and rarely a clear "right" choice, even with the benefits of hindsight. This is very different to when you take a test in school or college, where there is a right and a wrong answer and your ability to figure out which is which is a mark of success. Everyday problems are different; there may be "wrong" answers, but it's likely that there are also several "right" solutions.

Uncertainty is the only certainty. Accepting and adapting to this is crucial. The world can feel like an uncertain place, and it is. Unless we can learn to live with not being sure, it's very easy for us to become overwhelmed. What my research in psychology and neuroscience has taught me is that getting used to the intrinsic uncertainty of the world is essential for success: the people who thrive are those with the ability to accept and adapt to constant change and uncertainty.

The good news is that we can improve our ability to adapt. It takes practice and we often need to push ourselves outside our comfort zone. I managed to overcome my reluctance to speak in public and adapted over time to the demands of being an academic psychologist.

Harnessing the benefits of an agile mind—what I call "switch craft"— can be transformative. It's important to remember that we are active stewards of our own well-being, rather than passive victims of change and so we must actively manage our approach to life. Switch craft refers to those natural skills that are necessary to help us navigate a complex and unpredictable world. I have seen time and again how developing an agile mindset—the capacity to flex our thoughts, feelings, and actions—can transform our lives and bolster our resilience. In this book, I have brought insights from decades of work together in one place to uncover the mental talents we need to help us thrive during times of uncertainty as well as during more settled times. You will learn how to find ways to become more agile, find out what really matters to you, gain a deeper understanding of your emotions, and ultimately sustain your fulfilment, curiosity, and zest for life.

Maintaining a flexible mind allows us to thrive amid change. The first step on your switch craft journey is to accept that change and uncertainty are an inescapable part of life. Our lives will change, many times, sometimes for the better, sometimes for the worse. It is how we navigate those shifts that shape our present and our future happiness. If you are reluctant to change, or wary of trying out new things, this is something you need to work on—trust me, it will transform your life.

Agility is built into our DNA and supports our resilience. The good news is that nature has provided us with all the tools we need to become agile. While we may think that our own times are particularly unstable, most periods in history have been characterized by tremendous upheavals and uncertainties. People have always had to deal with wars, famines, floods, earthquakes, political upheavals, and pandemics. This is why we are actually intrinsically much more agile and resilient than we might think.

The key to resilience is our capacity to be agile and flexible in how we adapt to challenges and change. Our ancestors, alongside all the other creatures on planet earth, have always had to cope with a world that is constantly changing. We often lose this fluidity and become stuck in our ways as we grow older, but our built-in agility can still be released in a crisis or when we work hard to become more open to new ways of doing things.

Our brains have evolved to operate as "prediction machines." Think how frustrating it can be when a is missing in a sentence. Why? Because your brain predicted that a "word" should be present and its absence causes a surprise, or what's coded in the brain as a "prediction error." While we feel like we are reacting to what is happening around us, the way it actually works is that your brain constructs what is likely to happen next drawing on your rich experience of what's happened before. The latest science tells us that our every waking moment is dominated by predicting which actions we need to take next. As a result, our brain gives us a subtle heads-up about what's likely to happen from moment to moment, and this helps us to interpret our surroundings as well as the signals coming from within our own body. This continuous process gives each of us an exquisite biological capacity to adapt and respond, as long as we know how to harness it. Each prediction informs the body of what resources are needed, and the body then elegantly apportions its reserves to ensure that we are ready for whatever action is needed.

Our emotions are at the heart of our mental agility. Although these predictions generally occur outside our awareness, we can access them by what mindfulness teachers call "feeling tones." A surprising finding in the science of emotion is that each emotion does not have a specific feeling. Instead, what we feel is a general sense of pleasantness or unpleasantness—a feeling tone—and this informs us of what's going on around us before our conscious brain has had a chance to catch up. Feeling tones are a subtle window into our emotional life and provide us with a continuous readout of whether any action feels neutral, pleasant, or unpleasant. It is the feeling tone that gives a sense of urgency to every possible action and thought. In the noisy modern world, we often fail to listen to the signals coming from our own body and miss the wisdom that is contained within those feelings. This is why developing our emotional awareness and our intuition is so important. They help us to access the agile system that will help us to navigate all of the complexities in our everyday lives.

Paradoxically, our agile biology can also make us more rigid in how we behave. That's right: this agile, predictive system is also what makes us *reluctant* to change. While the system allows us to adapt, fast, it requires a lot of energy. Many of the actions that we predict actually never happen and this can be exhausting. Our tired brain can become preoccupied by worries and thoughts—in an ironic twist, the resulting unpleasant feeling tone infuses us with negativity where an inner critic can find evermore inventive ways of telling us that we are failing in some way, that we are not good enough. A vortex of negativity is released that keeps us more and more stuck in our ways as our brain tries to preserve energy and stick with old habits as often as possible.

This is why most of us inherently don't like change. I'm willing to bet that you have often resisted altering your well-established ways of doing and thinking about things. But ignoring change and doggedly trying to keep things the same will gradually, and inevitably, undermine your energy and vitality.

To stay agile and resilient we have to work at it.

An inflexible mind leads to anxiety and depression. In my decades of research in psychology and neuroscience, and in my coaching of countless business professionals and elite athletes, I have come to realize something as simple as it is extraordinary: an agile mind drastically improves your chances of success and happiness. But the flip side is also true: an inflexible mind fuels anxiety and stress and a "stickiness" that can torpedo your life.

The kernel of this understanding began early in my career, in a tiny testing cubicle where I obsessively measured the microsecond decisions made by our brains. I have always been fascinated by how our attention is captured by negative information. A spider on the wall, a creepy-crawly that runs across the floor, shocking news on the radio—all grab our attention. Being alert to danger is a hangover from the past, and we can only imagine how precarious life would have been for our ancestors. We all tend to focus on perceived threats—but for those who are anxious it is far worse.

For years scientists have grappled with the question of what happens in the brain, especially the anxious brain, when we are faced with threat. When I entered this field, the collective wisdom was that we have a threat-detection system deep in our brain that is constantly on the lookout for danger. When we become anxious this system goes into hypervigilant mode and for some people stays on high alert, even when they are safe. This is the essence of anxiety, it was thought, and means that we are constantly scanning our environment for potential danger. Lots of evidence fits well with this assumption.

I was never convinced that this "high alert" theory was the whole story. In some of my own studies I was noticing that the main problem for anxious people was not actually *scanning* for threat, rather it was in struggling to pull attention *away* from a threat once it had been detected. This difficulty in disengaging attention from a threat is very different from an enhanced ability to find threat in the first place.

What I call a "sticky" attentional system can lead to a rigid mind. It's like when you notice a spider and it's impossible not to keep checking back to see what it's doing. The same goes for our innermost thoughts, emotions, and actions. Once we think of a distressing thought it's often difficult to pull our mind away from it. This mental stickiness flows through our brain, leading to repetitive worry and rumination that keep us stuck, destroying our well-being and undermining our capacity to seize opportunities.

Self-help doesn't always help. In the developed world we have shelter, food, and a bewildering array of life-improving gadgets. What's more, decades of work in psychology labs around the globe have yielded many effective ways to help us thrive and reach our potential. Yet, many of us trudge through our daily routines rather than enjoying life. When I conduct workshops with successful businesspeople, the majority admit that they are neither as happy nor as fulfilled as they would like to be. What's gone wrong?

Endless self-help approaches claim to have the answer. It's important to be mindful, we are told, and to stay in the moment. Sometimes, we are advised to keep going no matter what, to be "gritty." Others tell us that adopting a "growth mindset" is key. These recommendations are backed up by solid science, and millions have improved their lives with these techniques. However, the complexity of the science is often oversimplified. The truth is that there is no "one size fits all" solution to dealing with life. Telling yourself to be mindful, or gritty, to banish your fixed mindset or to nurture positivity can be a little like telling a golfer to focus only on putting or to practice just the long shots; the match between your situation and the tool you are using gets lost in translation. There's little point in changing tack when grittiness is required, just as perseverance is useless when it's essential to make a change.

The more important factor in determining our happiness and success, I would argue, is knowing how and when to switch between different approaches. There is much evidence that we need a range of approaches on hand to deal with life's challenges. But range is not enough, we also need the agility to choose the right one for the right moment. This is the essence of switch craft.

The power of switch craft

Because the world is uncertain and complex, many different types of skills are required to deal with it. To return to our golf analogy, this is why a large number of different clubs are required to deal with the varying challenges of eighteen holes of golf. While I don't play myself, I have always found that golf is a perfect metaphor for life. Golf is peppered with problems—you may end up in a bunker, in the water, or even out in snake-infested woods. No matter where your ball lands you must deal with it in some way to get to the end point. And designers have been very inventive in designing golf clubs for every eventuality. It is similar in life. Finding the right approach for the moment is key. Learning several different ways to cope with challenges, and nurturing agility, so that you can choose the right approach for the right moment is the essence of thriving.

Building an agile mindset will help you to cope with change, and help you make better choices about how to approach any challenge or decision.

I am a cognitive psychologist and affective neuroscientist. I study the science of what makes us thrive at the Oxford Centre for Emotion and Affective Science (OCEAN)—a lab I founded and direct at the University of Oxford. We take into account people's genetic makeup, their brain functions, and what they tell us is important as we try to deepen our understanding of resilience and thriving. I also cofounded a company—Oxford Elite Performance—along with my husband, Kevin Dutton—another psychologist—to use cutting-edge psychology and neuroscience to help those in the sporting, business, and military elite reach their full potential. Having now coached many people to improve their performance in both sport and business, I have seen the benefits that improving agility can bring time and again. This has also dovetailed with what I am finding in the science lab. I have coined the term "switch craft" to illustrate this essential psychological talent, and the evidence for its effectiveness is growing all the time.

The four pillars of switch craft

There are four pillars of switch craft; each of them is important in its own right, but together they pack a real punch and will help you get through whatever life throws your way.

- **Mental agility:** The capacity to be agile and nimble in how you *think*, *act*, and *feel* so that you can navigate your way through all sorts of terrain, the rough as well as the smooth, and adapt well to changing circumstances. The science shows that agility is made up of four distinct components—what I call the "ABCD of agility": Adaptability, Balancing our life, Changing or challenging our perspective, and Developing our mental competence.
- **Self-awareness:** An ability to look inside yourself so that you can gain a deep self-understanding and appreciation of your core values and capacities. This will help you to become more aware of your hopes, dreams, and abilities.
- **Emotional awareness:** Part of self-awareness, but so important in our lives that it becomes a pillar on its own. Learning to accept and nurture *all* your emotions, those that feel bad as well as those that feel good, is vital. As is the ability to regulate your emotions and harness them in service of your values and goals rather than letting them boss you around.
- **Situational awareness:** This feeds off two of the other pillars, self-awareness and emotional awareness, but also incorporates the capacity to understand your immediate surroundings—to look *outside*—so that you gain a deep intuitive awareness of the context as well as your own "gut feelings." This mix of inner and outer awareness informs you as to how well you can operate in that environment.

Switch craft is like a compass that keeps you pointing in the right direction as you navigate your way through life. It can be learned and improved throughout your life. Whether it is coping with a difficult boss, managing a complex team, dealing with hyperactive children, resolving a dispute with a friend, or boosting your energy, your internal compass helps you choose the right strategy for the moment. If this compass is off even slightly, you can veer a long way from your course. Switch craft combines four vital psychological talents into a potent mental weapon to help you make the decision to stick or to switch to another approach, and to get that decision right more times than you get it wrong. Ultimately, that will help you to operate at the top of your game.

It is my hope that this book will bring inspiration from the frontiers of psychology and neuroscience to help you tackle the inevitable challenges that life will bring. Drawing on cutting-edge research in science, *Switch Craft* sets out a practical framework for how you can nurture the mental talents needed to live a successful, fulfilling, and resilient life. You will learn how to identify those thoughts and behaviors that are keeping you stuck in the past. You will learn the importance of nurturing a more open mind and how to make the adjustments and changes that allow you to become more agile. You will learn how to develop a deeper acceptance of uncertainty; it is only by loosening the shackles that are holding you back, by freeing those invisible patterns of thoughts and behaviors that fuel fear and anxiety, that you will be set free to find a more satisfying and fulfilling future.

Using this book

I would suggest that you make a journal part of your daily routine. There are lots of exercises and tests scattered throughout the book that will help you to become more flexible, learn more about yourself, regulate your emotions, develop your intuitive powers, and learn to prepare yourself mentally for any eventuality. Many people find that writing these exercises and thoughts down in a journal can be hugely helpful. Personally, I prefer an old-fashioned notebook, but you can start an electronic journal if that suits you better. Either way, your journal will allow you to keep track of how things are going, and the simple act of writing down some thoughts and exercises can be transformative.

The book is set out in five main parts. We start with the fundamentals of why switch craft is important, looking at the reality of change in our everyday lives and the importance of finding ways to manage the uncertainty and worry that can come with change. We explore the fascinating science showing that flexibility is a fundamental part of nature and finally we will see why agility is essential to building resilience.

We then move on to look more closely at each of the four pillars of switch craft. In Pillar 1 (Mental Agility) we explore the benefits of being agile; we examine the nuts and bolts of agility in the brain from an area of psychological research called "cognitive flexibility"; and finally, we explore the four key elements—the ABCD—of agility. In Pillar 2 (Self-Awareness), we discover why paying more attention to what your body is telling you is so important, and we take a deep dive into ways to find out who you really are and what really matters to you. Pillar 3 (Emotional Awareness) explores the nature of our emotions and how we can learn to understand and regulate them more effectively. The final building block of switch craft, Pillar 4 (Situational Awareness), examines the nature of our intuitive sense of the world, then we see how being exposed to many different life experiences can bolster our intuition and our understanding of the outside world.

At the end, I gather together some key principles of switch craft from across the book. My hope is that these switch craft skills will help you to learn to thrive and manage your well-being, especially in constantly changing and uncertain times.

Enjoy the journey!

THE FUNDAMENTALS

Why Switch Craft Matters

CHAPTER 1

ACCEPTING CHANGE AND ADAPTING TO IT

Usually, it is pitch-black. And deafening. The thup-thup-thup of rotor blades in the night air drowns out the sporadic swoosh of passing missiles. Lurching from side to side, it's impossible for the women and men crammed into the tiny space in the back to know how far the helicopter is from the ground, or even how far they are from their destination. "Two minutes," comes the call. Every member of the team then becomes absorbed in a private world of checking and double-checking. "Backpack secured," check; "head light in place and turned off," check; "jacket closed," check; "helmet secured," check. As the helicopter arcs toward the ground, a side door opens. The command to "Go, go, go" marks the exit of each team member as they are dropped at speed, one by one, from four feet above the ground.

Seconds later, the helicopter peels away so as not to alert anyone to their position and the team is running through the darkness to find the injured. It's the heat that first hits you, and the stench. And the smell of burning flesh is something you never forget.

Colonel Pete Mahoney commands the British Army's Medical Emergency Response Team (MERT). They work in small teams under the most difficult of conditions, operating much of the time in a dark helicopter. They have to keep their lights switched off so as not to attract enemy fire. Dropped into battlefields at great pace to treat the injured, they frequently come under intense gunfire as they make their way to the casualties. These teams typically comprise five to six people, including trauma surgeons, anesthetists, nurses, and paramedics, and at least two regular soldiers whose job is to protect the medical team.

Colonel Pete, a medical consultant, is often the most senior officer in the team and in overall charge. But a different group member will take control of the team at different points in the mission, depending on the nature of the situation. When they are first dropped on the ground, it's the soldiers who take the lead. Once the squad finds any casualties, one of the medical staff will then take over command and begin a systematic assessment of the nature of injuries—however, at any time the security detail can reassert command and order the team out of the area if it's judged to be too dangerous. Once an indication of priority of the casualties is given, the anesthetist then takes control and decides who can be sedated and brought back to the helicopter and who needs to be treated in situ. All this occurs on a live battlefield, often under heavy fire. These are rapidly moving situations that can change in seconds.

Less critically ill patients are allocated for treatment in a specific order, usually on the spot, and one of the nursing staff will typically take responsibility for this. The decision of how and when to bring the injured back to the relative safety of the helicopter is made by the soldiers, and command at that point is taken over by the helicopter pilot, who decides if it is safe to return and land in the location indicated by the staff on the ground. Then, when they arrive back on the helicopter, bringing with them the most seriously injured victims, a decision often has to be made about whether to operate straight away in the lurching almost-dark of the aircraft, or to wait until they get back to the field hospital. This is taken by Colonel Pete, in consultation with another senior medical officer.

It's hard to appreciate just how much mental agility these working conditions require. Colonel Pete is frequently in a position where he has to take orders from a much more junior member of the team. This is highly unusual in the military but the system optimizes the team's capacity to achieve their mission. It requires agility from all team members to cooperate and is highly effective.

While the conditions that MERT find themselves in are of course exceptional, they do reflect an extreme version of the constant changes and adaptations involved in everyday life. On any given day, our train could be late, the internet might crash, our child might come down with a fever. We could lose our job and be forced to relocate, a partner could say they don't love us anymore, a parent could die. The sooner we accept that change will happen, the sooner we will be on a path toward thriving.

Around us, political and social change still seems to be accelerating faster than we can keep up. The world looked askance at the vagaries of the Trump administration in the US, the implications of Brexit for Europe led to great uncertainties, and we were gripped by fear and insecurity as the coronavirus outbreak turned into a pandemic. It's hard to imagine that the iPhone was launched as recently as 2007; now the smartphone has spawned numerous companies such as Uber, Tinder, Airbnb, TikTok, Instagram, and many other industries that could not have existed without it. The coronavirus outbreak, meanwhile, supercharged the development of Zoom and other video-conferencing facilities that had been niche products prior to the pandemic.

At work, change should be "business as usual"

Despite this, in the business world, change is often thought of as an unpleasant cure for a problem, like surgery. I've seen this happen over and over again with many individuals and businesses. When a company is implementing a change, they often assume that change is temporary, with a beginning, middle, and end; it's something to be endured and needs specialists to take charge of it—in fact "change management" is now a thriving industry in its own right. But in reality, of course, change is not a one-off surgery; it is a continuous process and should be seen as a normal part of working life. Coping well with change depends on having the right mindset. Instead of creating a false divide between "change" and "business as usual," it's important to accept that change *is* business as usual.

At work, most of us perceive change as a threat. This is especially common when a company is going through a restructuring. Even when you see the changes being made as necessary, you may still be slipping out of your comfort zone. Of course, some changes might be threats, but when faced with something new and overwhelming you still need to take stock and assess what's happening with a cool head.

The thrive gauge

The thrive gauge helps to identify exactly what a change is and what are the positive and negative aspects of that change, by means of a traffic-light scale. The idea is to celebrate and maximize the greens, keep a careful watch on the yellows, but pay immediate attention to the reds, which may disrupt your aspirations. The key is to try to ensure that you are spending most of your time and energy on the activities that have a green light, while trying to think about working around the yellow- and red-light activities that might hold you back. It's well worth doing regularly.

1. **Step back and monitor where you are right at this moment.** Think about the nature of the change, perhaps write down two lists outlining some key advantages in one list and some

disadvantages in the other. Then make a separate list of your short-, medium-, and long-term professional goals.

2. **Use a traffic-light system to get perspective on how the change might affect your own personal goals.** Give yourself a "red light" for the elements of the change that might get in your way, a "yellow light" for the danger signals, and a "green light" for things that could work well for your goals. For instance, if your company is moving from individual offices to open-plan, say, you might be nervous about feeling that you will always be "on show," or that the ambient noise will affect your proficiency, or that confidential discussions with clients might be overheard. Each of these might get a "red" or a "yellow" light, whereas a "green" light might be given to the opportunities for collaboration or striking up creative conversation with others.

I was coaching David, a senior manager at a property development company, when his bosses announced that they were merging two large teams of people. David led the commercial team, and he was told they were now to join forces with the residential team. This caused David some excitement but mostly great concern. His "red light" was the fear that he would lose the sense of fun, intimacy, and strong teamwork that he enjoyed with his current team. With a much larger group, he feared that keeping the same ethos would be impossible. A "yellow light" was that he would need to divide his time between two different physical locations and this might eat into his family time. However, the big "green light" was that he would likely be appointed to lead the larger, merged team, and this would be a serious step toward his leadership ambitions.

David used this light system to help him work his way through the change. He organized several social events with smaller groups that combined members of the former "residential" team along with members of the more familiar "commercial" team. This helped the previously separate teams get to know each other and kept the sense of fun and teamwork that David enjoyed. He did now have to divide his time between two locations, but he tried to manage this as best he could so that he only had to stay away from home one night each week. He also enrolled in some short leadership courses to keep him on track for his "green light" ambitions of moving up the leadership hierarchy in his company.

Changes and transitions

The self-help writer and life coach William Bridges makes an important distinction between change and transition. *Change* is the external events that happen all through our lives. *Transition* is different. Transition is the subtle internal reorientation and self-definition that are necessary in order to deal well with changes in your life. "Without a transition," as Bridges tells us, "a change is just a rearrangement of the furniture." Many of us make detailed preparations to get ready for a big change in our lives such as the arrival of a new baby or the move to a new job, but we rarely think much about preparing for the internal transition, and this can really blindside us.

I worked with a top athlete—let's call him Harry—who'd had a highly successful career in sport and retired at the age of thirty. He was plagued by injuries and was finding it hard to maintain the level of intense training required. He knew retirement was the right decision; he was clearly past his peak and realized that it was time to move on to something else, but nevertheless he told me he was really struggling with his new identity of "ex-sports-star."

In the year before retirement, he had gone to great lengths to prepare for the change. He signed up with speakers," agencies to give motivational talks, he spoke to TV and radio companies about becoming a sports commentator, he enrolled in a coaching course so that he might have the option of becoming a coach. To begin with, all of this was going well and his commentating work especially began to pick up. But in the year after he stopped competing, he was becoming more and more unhappy. At first, he thought this was to do with a lack of structure. He had gone from three training sessions each day to nothing. He started to get up early and go for a five-mile run every morning at the same time; he would then shave and have breakfast. These simple rituals helped a lot. However, boredom during the day was still an issue and he had begun to drink heavily and was now out almost every night, which was causing problems in his relationship with his wife. He was becoming unreliable, and a major agency dropped him from their books.

The problem, he told me, was when he looked in the mirror. "Who do I see?" he asked himself. No longer "champion" or "sports star," so who? We realized this was the nub of the problem. He no longer had a clear identity; while he had prepared well for the *change*, he had not prepared for the *transition* from much-feted athlete to ex-- sports-star. He was essentially stuck between identities.

Every new beginning starts with an ending

The process of transition involves first letting go of the old situation, then suffering the bewilderment and confusion of the in-between state, and finally emerging to start again in the new situation. To transition well, it is essential to go through each of these three processes. I encouraged Harry to go right back to the beginning—the ending of his career—and he spent several months really thinking about what it meant to give up professional sport, which had been his life since he was about ten years old. It was effectively a type of grieving process. Thankfully, he is now much happier in his new identity of sports commentator and mentor to young people just coming into the sport. Together we came up with four general principles that helped Harry to navigate this difficult transition.

1. **Respect the process**: growth occurs at its own pace, and you should not try to force it.
2. **An internal shift in identity is essential**: you can only adapt to new circumstances by changing inside.
3. **Accept yourself** for who you are and what the process of change entails.
4. **Lower your expectations** about what you can and can't do during this period.

The fertile void

Change in our working and personal life is to be expected. And as we have seen, these changes may challenge our sense of identity and require an internal transition to cope well with the change, whether we know it is coming or not. A man who has been married for many years may really struggle if his wife leaves him and he finds himself single again. This type of personal change has profound implications for our lives. So, how can we help ourselves to navigate these types of changes?

A large body of research has shown us that while change can be scary, especially when we are already feeling anxious and apprehensive, we can learn to cope with it. What you need is time, and self-compassion, to allow yourself to disconnect from your previous plans and gradually begin to accept and open yourself to new goals and possibilities. The German psychoanalyst Fritz Perls coined the term "the fertile void" to refer to this difficult space between the end of one thing and the beginning of another.

On a much faster timescale, my own laboratory experiments reveal disruptions and delays—"switch costs"—when people shift between very simple mental sets like going from "categorize a digit as odd or even" to "categorize a digit as greater or less than 3." An interval is required to disengage from one mental set (odd/even) before you can reengage with the next (larger or smaller than 3). So imagine how much more time and effort is required when you are switching from a major aspect of your identity: from "happily married," say, to "divorced" or "single" or "widowed." Allowing yourself a "fertile void" is essential. And, as Perl's phrase implies, this is not just dead time, but a productive and essential period to disconnect and then gradually reengage.

Major transitions take time, and effort. Losing your job, a relationship ending, a close friend dying—all will stop you in your tracks and force you to reevaluate your life and your goals. Any change will confront you with uncomfortable questions and challenge many things that you may have taken for granted. Our natural temptation is to try to avoid the pain of major change, perhaps by becoming absorbed with work or even by anesthetizing ourselves by drinking or taking drugs. But it is important to allow yourself time to experience the pain and distress and to take time to get used to the new situation. This is equally important when the change is a positive one such as a new relationship, moving to a new home, or changing to a new career. Rather than moving straight from one job to another, for instance, try to take some time out—it may be as simple as going away for a weekend—to ensure a bit of space between jobs. Even within a day, take time out and perhaps go for a walk, meditate, meet a friend for a chat—anything to insert a natural gap.

Your brain needs time to adapt, no matter what life event you are going through. Dealing with change requires multiple tiny adjustments as you move from the initial surprise to gradual acceptance and adaptation to the new reality. Here are some steps I worked on with a couple who were making a major move of both career and country.

1. **Designate a set part of the day to be alone.** Dedicate this time exclusively to sitting quietly so that you can get in touch with your innermost feelings.

2. **Take the time to list all the ways in which your new life will be different.** I asked them to make a point of trying to think of all the changes that would happen when they left their familiar place, and the effects this would have on other areas of life. It's important to go into detail. For example, they wrote things like: "We won't have as much income as usual for a few months so we'll have to be careful about our budget," and "We will have to make a real effort to meet new people every week because no one knows us" or "We will need to take the time to explore the different neighborhoods and decide where we might want to live." Prioritize these questions and concerns in terms of what is most important to you, which if answered, would reduce a lot of your anxiety. Knowing the kind of place you could afford to live is important and so perhaps spend an hour or so each week having a look at properties and prices in different areas online. This gives a sense of what to expect and helps to replace feelings of doubt with feelings of excitement. There are also deeper questions that are important to ask: "Is my current role important for my identity?" "Will people look at me in a different light if I leave?"

3. **Take the time to mourn losses.** Expect and accept signs of grieving and don't confuse them with low morale or failure. The couple I was working with knew that they would miss places and people and felt very sad about it. And that's OK. It's natural to feel sad, scared, depressed, and perhaps confused. They went through a period where they constantly wondered whether they were making the right decision. It's important to allow yourself time to go through this discomfort. Don't feel that you have to make the feelings go away.

4. **Define what's over, and what isn't.** It's not all sadness and tears. Look at the big list of changes and group them into themes. Look out for interesting opportunities that might tell you more about yourself. Many of your themes will probably revolve a lot around fears and doubts: loss of your comforting rituals,

daily routines, or aspects of your identity that have become familiar. Some things might be gone forever because they are tied to a time and place, but not everything has to end permanently. Identify those things that can be transformed or adapted to your new situation.

Initiating change

When I was twenty-six years old, I realized I had to quit smoking. For a long time, I had been in blissful denial, and shut down the possibility or even the advisability of quitting. But I was playing a lot of tennis at a competitive level, and toward the end of long matches I began losing points simply because I was running out of steam. Smoking was undermining my endurance and this scared me.

I took a good hard look at myself—what I would now call a "situational analysis." Why was I doing something that was so obviously bad for me? There was no good reason to continue smoking. I had tried, and failed, many times to cut smoking out completely. And so, my initial plan was to only smoke after lunchtime. This was hard at first, but gradually I got to the point where I could get to lunchtime without thinking too much about smoking. This was a fragile success and I was all too aware that any stress or unexpected event would send me scurrying back to the cigarettes in the morning. But gradually I pushed the "smoking ban" to after 3pm, then after 6pm, then after 8pm, and eventually no cigarettes at all.

There were several lapses. On many occasions, usually at parties, my mind would play inventive tricks to convince me that not smoking was ridiculous. Everyone else was enjoying it—so why shouldn't I? It wasn't really that bad for me. What difference was one cigarette going to make? All of these thoughts were constantly reverberating around my head, weakening my resolve. There is nothing more creative, and convincing, than the mind of an addict.

I began to accept that a lapse was a lapse, nothing more, nothing less. The important thing was to start again right away and, this time, try not to slip up. About eighteen months later I gave up for good and have not touched a cigarette since. The important point came when I began to see myself as a nonsmoker. My identity had changed and, as we saw with Harry, this transition is vital to help you maintain a change in your life.

Dealing with changes that happen to us is one thing. But, of course, there are also times when we seek out a change. This can be just as difficult, if not more difficult, as dealing with changes that have been imposed upon us.

The five key stages to making a change in your life

Extensive research on quitting smoking has shown there are five key stages, which are applicable for any change you want to implement in your life—whether it's getting fit, losing weight, or changing career. I recognize all of them in my own attempts. The first two stages are when you are simply not ready to change, and the mistake many people make is not figuring out *what* it is they want to change and *why*. You then enter a period where you are aware of the benefits of change but also a bit scared of the downsides. Weighing these pros and cons can take some time swinging between procrastinating to deciding that it's time to change. Only now are you ready to develop a plan and take your first action. You might join a Zumba class, quit carbs, or talk to people about different career options. The final stage is about maintenance. How can you ensure that you will be able to stick with your new actions and goals? This might mean not having cigarettes in the house, surrounding yourself with people who are regular gym-goers, or simply scheduling regular exercise sessions in your calendar.

Here's an outline of each of the crucial stages to making a change in your life.

Stage 1: Pre-contemplation

Conduct a "situational analysis." Write down a list of things that are working well for you and that you want to keep in your life: these could include certain friends, habits, or hobbies. Now, write down a list of things that are not so good, things that you might like to change: these could include certain friends, habits such as smoking or drinking too much alcohol, or perhaps you would like to improve your sleep. At this stage, don't make any big decisions or plans. Simply make one list of those things about your life that you are really happy about and another list of those things that you are not so sure about.

Stage 2: Contemplation

Now give some serious consideration to anything you might want to change, stop, or start, and make a list of the benefits and costs. Then, make a clear decision. I want to stop smoking; I want to stop drinking alcohol during the week; I want to exercise three times a week. Make the decision specific—rather than, "I want to lose weight," give a specific target, "I want to lose seven pounds over three months." Then, tell someone. There is significant evidence that informing others of our intentions makes it much more likely that we will stick with these good intentions.

Stage 3: Preparation

Now is the time to start planning and thinking about your new routine. For example, if you want to lose weight, make a list of the times and situations you are most likely to overeat or least likely to exercise. Outline your triggers—it might be with your morning coffee, talking on the phone, or when the weather is cold. Once you are aware of these triggers, you can think about alternative coping strategies. Are there any changes in your environment or behavior you should make, even for a while? Sometimes, simply having a set time to exercise can be helpful. Schedule a gym session or a run at a specific time and, regardless of the weather or how you feel, make a pact with yourself that you will just get out and do it. Once again, tell other people about your intentions and make a list of supportive people you might use to help you. This stage is all about planning how you are going to implement your change.

Stage 4: Action

Now is the time to put your plan into action. All the plans in the world are useless if you don't follow up on them. So, once you have your plan—"I will go to the gym at 5pm," say—then don't make excuses, just do it without thinking. If you want to start getting up earlier in the morning, then set your alarm and get out of bed as soon as it goes off—no snooze button, simply get up. Over time, you will find that you don't have to think about your plans too much; you'll just do them almost automatically. (It's amazing how often overthinking can upend our best-laid plans.) Finally, don't forget to reward yourself for success. Schedule in a reward, perhaps each week; just make sure it isn't something that may trigger you to go back to your old ways.

Stage 5: Maintenance and relapse

You will almost certainly relapse from time to time. If you do, don't beat yourself up about it, instead show some self-compassion. What you are trying to do is hard. Turn the relapse into a valuable lesson, from a stumbling block into a stepping-stone. Keep a record, outlining your successes and failures, in your journal. Go over the situations, feelings, and triggers that led up to the slip. For me, when quitting smoking these were usually when I was feeling tired and stressed at work, or when I was tired and went out to relax with friends. Fatigue was a clear danger signal and so I tried to make sure that I got enough sleep and avoided socializing when I was feeling tired. Ask yourself what you could have done differently. What didn't you consider? What changes can you make to keep a slip like this from happening again?

Watch out for rigidity in your approach to life

Persistence is often good, but not always. If you are constantly failing to make a change and having to start again, you might want to consider a different approach. You need to watch out for the creep of what we can call "mental arthritis" in sticking relentlessly to plans that are not working. By mental arthritis I simply mean inflexibility in being able to act or think in a way that is most appropriate for the situation.

In my case, attempting to cut smoking completely failed every time. It was too big a thing to tackle all at once. But, once I scheduled certain times I was allowed to smoke, and broke the "stop smoking" project down into bite-size pieces, it became much easier. Moving from two cigarettes a day to none was much easier than going from a whole packet to zero. Often it is about small steps that are not difficult in themselves but gradually build up to a range of habits that enhance, rather than detract from, your life. Sometimes we also get stuck at the transition stage. To begin with I was thinking of myself as trying to be an "ex-smoker," but I realized I actually needed to make a more radical shift in identity, to "nonsmoker."

Variety is the key to coping with change. It's important to constantly remind ourselves that there is rarely a single solution to most of life's problems. The idea that there are simple secrets to change that you can implement easily is appealing but misleading. Life is much more complex and the problems we grapple with throughout our lives require many different solutions and the agility to try out different approaches.

The great religions have always known this. Organizations that you might imagine have very entrenched, and rigid, rules are actually surprisingly flexible in their approach to life. Ever wondered, for instance, why there is such a bewildering variety of yoga practices? In the Hindu tradition alone we have karma yoga, bhakti yoga, jnana yoga, and saranagati. Different versions are thought to help us achieve somewhat different spiritual goals. The anecdotes attributed to Krishna, the Hindu god of compassion and love, make it clear that there is no single path to reach awareness and every one of us must find our own way—a degree of flexibility is allowed.

Buddha also allowed for what he called "ethical flexibility," which was the suggestion that while the ancient texts provide us with "wise counsel" in the form of guidelines for living a good life, these were not to be treated as rigid commandments or rules. Instead, a person is required to use their own judgment in terms of how they should behave rather than being obliged to act according to unbending values that cannot be transgressed. Similar ideas play a central role in Islamic faith, with a small number of forbidden activities at one end and a small number of obligations at the other. Most human activities occur somewhere between these two extremes and, again using their own judgment and conscience, people are encouraged to show individuality in terms of approach and lifestyle to develop a singular and yet legitimate Islamic practice.

Christians are also encouraged to develop agility. The founder of modern-day Christianity, St. Paul, acknowledges the importance of agility when he admits: "I have become all things to all people so that by all possible means I might save some."

The American actor and martial arts fighter, Bruce Lee, has championed related principles. On considering life, Lee concludes that we often make the mistake of trying to force the world to adjust to us, rather than being agile and adaptable and aiming to deal with the situation in the most appropriate way possible. "Be like water," Lee tells us. Rather than sticking with rigid beliefs and actions, we should be more like water and find our way through the cracks. To truly express ourselves as a human being, Lee says, we must resist affecting a style that we stick to because a "style is a crystallization" when what is needed is a process of continual growth. "Running water never grows stale, so you've got to just keep on flowing," he says, returning to his favorite metaphor.

Switch craft is the capacity to cope with the big changes and transitions in life and finding ways to deal well with unexpected twists and turns. These skills are essential because uncertainty and change are facts of life. So, learning to not just accept but also to embrace change is an important preliminary step in your switch craft journey.

Chapter Summary

- Accepting that things won't stay the same is essential for thriving.
- It's important to allow yourself a "fertile void" when transitioning to a major change.
- To instigate change in your life, you need to move through five stages:
  - *Pre-contemplation*—deciding that a change would be a good idea
  - *Contemplation*—thinking how you might go about changing
  - *Preparation*—coming up with a plan of how you will change
  - *Action*—putting your plan into action
  - *Maintenance and relapse*—thinking about how you can keep the new behavior going
- To manage a constantly changing world you must acknowledge that a variety of different approaches are needed. There is almost never a "one size fits all" solution to life's problems.

CHAPTER 2

MANAGING UNCERTAINTY AND WORRY

Sometimes life throws us a surprise that shakes us to our core. Panic rises, and it can feel like standing on the edge of a chasm knowing just how easy it would be to tumble in. Whether it is a war, a tsunami, or a pandemic, a crisis brings our own precariousness and vulnerability to the forefront of our minds.

Uncertainty is a normal part of life at the best of times and never more so than during the coronavirus pandemic that began in 2020. As the virus spread around the world, question after question emerged: When will we be able to travel again? Will there be further lockdowns? Will schools close? Will we be able to develop a vaccine? Will my business survive? It became evident that there were no clear answers. The uncertainty facing our way of life was palpable. Countries shut their borders; planes were grounded; restaurants, bars, and clubs closed; many people had to work from home, some were placed on temporary furlough or lost their jobs completely, while others had to do something completely different.

The power of rituals

Rituals can bring some structure and order to a chaotic and uncertain world. During the coronavirus pandemic, social distancing measures led to the cancellation of many events such as concerts, sporting competitions, and memorial celebrations, as well as more personal occasions such as christenings, bar or bat mitzvahs, weddings, and funerals. I spent a lot of my time on Zoom calls advising companies and their staff about how to optimize their well-being during these unprecedented times. A key piece of advice that was effective was suggesting they introduce some rituals and structure to their time. Perhaps getting up an hour early and having a walk or a run or a yoga session before breakfast. Taking the time to call a friend every week. Scheduling a time each day where the phone is turned off so that you can do some reading or listen to music.

These small rituals can have a surprisingly big impact in helping you to deal with the uncertainty that is an inevitable part of life. The reason they work is connected to the fact that our brains are prediction-making machines. Because your brain cannot possibly obtain all the information necessary to make completely accurate predictions, it regularly gets these predictions wrong. For instance, your brain might predict the height of a step as you are walking up a flight of stairs and you move your foot appropriately; but if the calculation is out by just an inch, you must make a quick adjustment. This is coded as an "error signal" by the brain and is stored away to make sure that you get it right next time around. Your brain learns, quite literally, through its mistakes.

Although this is all going on below your conscious radar, if the brain is signaling numerous prediction errors—as happens when our usual routines are disrupted—you will notice feelings of apprehension and anxiety. Your brain is a kind of "uncertainty detector": as uncertainty increases, your vigilance and feelings of anxiety increase, while when you are in a more controllable situation you can relax.

Rituals are effective because they give your brain the opportunity to make some more stable predictions, and some space to get perspective on everything else.

The consequences of the pandemic ticked all the boxes of what we know can trigger strong feelings of uncertainty for all of us. When our whole world is thrown into chaos, we are faced with a much more profound sense of uncertainty than the type of specific changes that we explored in the previous chapter. This broad sense of uncertainty can fuel deep-seated anxieties and worries.

What triggers uncertainty?

Research shows us that there are two broad types of situations that are very likely to cause a sense of uncertainty:

- **New situations:** When you are in a new environment such as starting a new job with people you do not know, or when you arrive in an unfamiliar country.
- **Ambiguous and unpredictable situations:** Life is full of ambiguity where it is difficult to know whether there will be a happy or a bad ending. You have a mild pain, say, that may indicate something very minor or may be a sign of a more serious illness; somebody might make a vague comment such as "I hear there are changes afoot in the company," which could indicate good news, such as a new acquisition or series of promotions, or could be bad news like forthcoming redundancies or the closure of a branch. Watching a football match where the score is close and the outcome is unclear; a negotiation with a potential investor for your company where you think she may invest but it is still uncertain; an interview where you are unsure whether you will be offered the job; or when you are waiting for the

results of an exam that will have important implications for your future.

These situations can cause our levels of anxiety and stress to skyrocket. We can think of uncertainty as being like an allergy: even a small amount can cause a bad reaction, while larger amounts cause even stronger reactions. As human beings, we crave security, and that is why all of us are intolerant of uncertainty to some extent. This is partly because ambiguity and uncertainty can be highly energy-consuming for our brains. Remember, our brain likes being able to predict the future, so we have a natural drive to reduce any ambiguity by trying to impose certainty and predictability on our lives.

But there will always be an infinite number of "what if" scenarios to cycle through. You might lose your job. You might be diagnosed with a fatal disease. World War III might break out. Your child might be abducted in the playground. We might all be part of a giant matrix game orchestrated by a distant alien civilization . . . And the trouble is, we cannot *definitively* rule out any of these possibilities, even those involving aliens. So, whether we like it or not, uncertainty is something that we must accept.

How tolerant of uncertainty are you?

What in psychology we formally call "intolerance to uncertainty" is effectively an index of how much we fear the unknown. Each one of us has different tolerance levels for uncertainty, but this is not "hard-wired." Our tolerance can also fluctuate depending upon how we are feeling. When we feel under threat, we become more intolerant; when we feel relaxed and safe, our level of discomfort with uncertainty goes down. The more you struggle with uncertainty the more you are likely to try to avoid the natural ambiguity of life rather than develop ways to cope with it. The good news is that it is possible to change your tolerance level, especially if it is having a negative impact on your enjoyment of life.

The following questions will give you an idea of how you measure up in terms of feeling uncomfortable with uncertainty. Answer each question honestly and for each give yourself a score of 1 to 5 according to how much each statement is typical of you (1 = not at all like me, 2 = a little like me, 3 = somewhat like me, 4 = very like me, 5 = completely like me); then add up your total score.

1. I really don't like surprises.
2. I get frustrated if I don't have all of the information that I need.
3. There are many things I don't do if I am unsure about them.
4. I always try to plan ahead to avoid unexpected things happening.
5. Even small things that happen unexpectedly can spoil a well-planned day.
6. I sometimes cannot do things because I am paralyzed by uncertainty.
7. I always want to know what is going to happen to me in the future.
8. I don't function very well when I'm uncertain.
9. If I have any doubts about something I find it very difficult to act.
10. I try to avoid all uncertain situations.

> o A score between 10 and 12 is *Very Low*
> o A score between 13 and 15 is *Low*
> o A score between 16 and 28 is *Average*
> o A score between 29 and 45 is *High*
> o A score between 46 and 50 is *Very High*

A low score shows that you have a high tolerance for ambiguity—it is likely that you are curious about the unknown, and you are quite happy to take on new and possibly inconsistent information. On the other hand, intolerance to uncertainty, indicated by a higher score, leads to a disproportionate need to feel safe and secure alongside a tendency to worry, which in turn of course generates anxiety and stress. When unsure, we try to figure out how we can stay safe and engage in what are called "safety behaviors." Safety behaviors are any of those things that provide reassurance and reduce uncertainty—so, phoning your teenager if you are worried about where they are, or checking menus in a restaurant before you go there to see what they have on offer. All of us can work on ways to decrease our need for certainty using the exercises in this chapter and become a bit more used to simply not knowing.

Seeking safety and certainty is not in itself a bad thing, but it can easily become obsessive and lead to an escalation of anxiety. Nowadays, of course, most of us carry around a smartphone, which is effectively a "certainty-seeking device." If I'm unsure of the capital of Bangladesh, my phone can tell me it's Dhaka in less than a second. If I don't know whether my friends are nearby, I can text them and get an immediate reply or even track them via an app. If I'm wondering where the nearest pizza restaurant is, I can find out instantly.

Psychologists have been researching whether these devices may be leading to higher intolerance of uncertainty and anxiety. One study analyzed data from the United States between 1999 and 2014, during which time there was a large increase in mobile phone use and a coinciding marked increase in measured intolerance to uncertainty. Of course, many other things would have changed in society during this time period that might have had an impact, so we need to be somewhat cautious about how we interpret this type of data. Nevertheless, we know that some exposure to uncertainty is a good thing, and since mobile phones act as instant safety cues, it seems likely they may effectively reduce our everyday exposure to uncertainty, lowering our tolerance levels and increasing anxiety.

There's little doubt that a certain degree of planning and trying to establish predictability in your life is useful and a good way of managing stress. However, there is a fine line between forward planning that is helpful and a degree of trying to eliminate future uncertainty that becomes unworkable. If we are uncomfortable with even a small degree of uncertainty, our increasing attempts to find safety and certainty can become a problem. Common behaviors include:

- Regularly seeking reassurance
- Searching out information
- Excessive list-making
- Refusing to delegate and insisting on doing everything yourself
- Constantly double-checking
- Overpreparing
- Aiming for perfection
- Procrastination
- Avoidance of novel or spontaneous situations

Fear of uncertainty is the fuel that drives our worries

The reality is that there are almost no circumstances—if any—in which we can find complete certainty. And so, logically, not being 100 percent certain is something that we simply must get used to. Yet, just as we don't like change, we also don't like ambiguity. This is why worrying is a common by-product of struggling to tolerate uncertainty. When we are uncertain, our threat-detection system can go on high alert, which of course is a perfect way to begin and maintain a spiral of worry and anxiety.

Worry is a way of coping

When we feel unsure, worrying is often an attempt to gain some control of the future. It is a way of mentally preparing for a bad outcome.

Worry that leads to action can be productive, but often it is unproductive. We imagine that if we agonize over every "worst-case scenario" and aspect of a problem, we'll find a solution. Unfortunately, this just doesn't work. Constant worry does not give us any more control over events; it simply saps us of energy and destroys our vitality.

It is often not specific topics that trigger our worries, but rather the uncertainty that surrounds them. In the morning you may be worried about getting the children to school on time, in the afternoon you may be worrying about the health of your family, while by evening you may be fretting about whether you should replace your aging washing machine. It's not the specific things that you are worried about, it's the uncertainty.

Intolerance of uncertainty affects our ability to challenge ourselves

When we encounter a problem in our personal or our work lives, part of the struggle is that we don't know what will happen next. But, if you think about it, every decision *has* to be made in a sphere of uncertainty. People who are cool with uncertainty often enjoy ambiguous situations because they have a thinking style that is open to new and challenging information, whereas those who are very uncomfortable with it tend to have a more closed thinking style that seeks familiar and predictable situations.

Some of us cope with this uncertainty in decision-making by choosing a solution that is either very easy or very difficult. The easy option is appealing because it is, well, easy, but the difficult option can also be tempting because it will not damage your self-worth if you fail. One of the young track athletes I work with was struggling with this. She would either take on the impossible or set her sights too low. When choosing training partners, or during races, Katie would often measure herself against someone much slower than her so she could easily stay out in front. At other times she would train with people who were significantly faster and more experienced than her. And while this did help her to achieve faster times, we realized that she was still holding back, because no one, including herself, expected her to beat the more experienced runners. Katie admitted to me that she was most nervous about competing with people who were very close to her own ability—the situation where the outcome is most uncertain.

We agreed that she should make a step change, and start regularly pitting herself against runners of a similar ability. As we talked through this, it also dawned on Katie that this was why she often became very nervous and then underperformed in important events such as national trials, where she couldn't avoid her peers. She worked on her intolerance of uncertainty for several months using some of the methods I will discuss later in this chapter, but mainly by exposing herself to more and more training sessions and races where the outcome was genuinely uncertain. She entered races in different distances from her own to see how she did and selected new training partners who were closer to her ability. As the uncertainty became more and more familiar, she actually began to enjoy it. Thankfully, her anxiety subsided dramatically and her performance against her peers at important events improved, and she is now regularly competing for national team selection. Many of us are familiar with the idea that you learn more from failures than successes, but perhaps less well known is that most learning and improvement will come in the middle ground between the very easy and the very difficult.

Fear of uncertainty pushes us to make hasty, and familiar, decisions

Numerous research studies tell us that uncertainty pushes us toward the familiar rather than considering a range of options. We see this in the way people react to a terrorist threat, which often results in a surge in approval and trust for the current political leadership, a desire to attend religious services, and a greater tendency to display symbols such as the national flag. We attempt to cope with our intense feelings by trying to find some stability, certainty, or familiarity. In one intriguing set of studies, a group of American students were exposed to a video of a terrorist attack. Reminding the students of these attacks led them to make dramatically more rapid decisions on totally unrelated issues. And the more uncertain they felt, the more likely they were to choose political leaders described as "decisive" rather than "open-minded."

When we are uncertain, stressed, or tired, our brain will crave firm answers. Our need for closure on complex issues is strong because the motivation is often to reduce the degree of uncertainty felt, rather than reaching the best answer. Of course, it's not always possible, but it is generally better to make important decisions, especially where the outcomes are uncertain, when you are feeling relaxed and secure.

Becoming more comfortable with discomfort

Therapists often use the metaphor of a sailing boat to capture this idea. Imagine that you are journeying through life in a small sailing boat, and you are the captain. As you steer your way through the waters, some calm, some choppy, waves occasionally slop over the side, wetting your feet and making them cold. It's not life-threatening and you know it won't sink the boat, but it's uncomfortable. There's a small bucket in the boat that you use to bail out the water. The more water continues to come in, the more you use the bucket to bail out the water. As you are bailing and trying to clear the water, you have a look at what's been happening to your boat. Is it still heading in the right direction? Have you been paying more attention to bailing than sailing?

Now you have a closer look at your bucket, and you see that it's full of holes. You have been using the wrong tool for the job, and meanwhile your boat has been drifting. Would you rather be in a dry boat that is drifting aimlessly, or to put up with a little water and cold feet and head in the right direction? Worry is the leaky bucket in our story. It is the wrong tool for dealing with uncertainty and causes us to avoid confronting the issues that it is necessary to face. Increasing our tolerance of discomfort and uncertainty can be transformative. It helps all of us, including our political leaders—if we give them a chance—to make better and more informed decisions and achieve our goals.

You don't need to always feel good to perform well and achieve what you want to achieve. And that is worth reflecting on. Sometimes we need to accept uncertainty and stress, and, frankly, feeling crappy, and just get on with it. Those who thrive do not avoid negative thoughts and feelings and being uncertain. They accept them as a normal part of life. Success in psychotherapy, for instance, is not when people feel happy and energized—although that's great if it happens—it typically comes when people learn to live more comfortably with uncertainty and their negative feelings.

Building up exposure to uncertainty

Catastrophizing is when you jump to the worst possible conclusion, often with vivid images of your fears unfolding. A businesswoman I was coaching, Alexa, had real issues when her partner, Ahmad, did not call her at regular time points. It was becoming such a problem that her husband's anxiety was also escalating. Anytime a meeting ran over or Ahmad got stuck in traffic and could not call her, he began to get flustered because he knew she would be panicking. As the uncertainty continued, she would become more and more anxious and would engage in a range of safety behaviors such as seeking constant reassurance that he was OK by texting him repeatedly, texting colleagues or friends that she thought he might be with, or even checking the local news for traffic accidents. None of these activities helped, instead they typically made matters worse; and when they did finally make contact, he would be greeted with her anger that his lack of communication was causing so much angst.

The three of us worked together on a plan to help Alexa become more comfortable with being uncertain and to help her understand that the unpredictability of life is not always a bad thing. We arranged some pleasant surprises—Ahmad unexpectedly telephoning just to see how she was, or dropping around to her office with flowers and taking her out for lunch. On other occasions, she agreed that he would miss one of his regular calls and she promised not to try to get in touch. Alexa found this extremely distressing, but she eventually began to recognize that her worry and her reassurance-seeking was making the problem worse. It's always worth taking a moment if you find yourself catastrophizing and considering any advantages that it offers. Usually, you will struggle to find any.

I encouraged Alexa to think deeply about her negative beliefs around uncertainty and to test them. One of these beliefs was, "When I'm uncertain about how Ahmad is, I won't be able to focus on anything else." I asked her to put this to the test by not getting in touch with him for an hour and focusing on a meeting, or on completing a project. She was not particularly comfortable with this, but realized that it was possible and that nothing bad happened.

The power of "behavioral experiments"

The exercises Alexa and I worked on are examples of what psychologists call "behavioral experiments"—these are activities that test our beliefs and expectations. They work by exposing people to small "doses" of uncertainty and by exploring the usefulness of worry, among other things. By targeting beliefs through actions, you can test out your most feared expectations. They are very effective at bringing about change in how you think about uncertainty and softening your conviction that uncertain situations are always negative.

Consider planning some behavioral experiments of your own, focusing on areas of your life where you are less tolerant of uncertainty. Start small: if you are scared to attempt reconnecting with a relative after a family feud where there has been a lot of bitterness and hostility and no contact for ten years . . . perhaps leave that for now. Instead you could reach out to a friend or acquaintance who you have not seen for a while to test how that feels. Then, you could tackle a mildly difficult conversation—such as bargaining for a discount in a shop or hotel. The idea is to start with a small degree of uncertainty, and then gradually expose yourself to more and more degrees of uncertainty as you become more comfortable.

Are you constantly scrolling through social media to keep up with the news? Scale this down to checking once every half hour. Then, once you've got used to this, have a check every hour, and so on. Eventually, try to check just once or twice a day, ideally at scheduled times. Give yourself breaks—go out for an hour and don't take your phone with you, or simply turn it off for a couple of hours. (I suggested this to a group of teenagers in a school talk recently, and they were absolutely horrified.) Don't avoid these activities completely—the idea is to gain some control over your actions and get more comfortable with not knowing.

Approach these experiments with curiosity. Becoming more open and interested about "what might happen" takes the sting out of uncertainty and will allow you to consider the positive things that might happen rather than always assuming that these situations will end badly. Write down the outcomes you are expecting: Is this negative, positive, or neutral? Then, write down the actual outcome: Was it negative, positive, or neutral? If it was a bad outcome, how did you cope? Were you able to deal with the situation? Could you think on your feet? Was there something else that you could have done? This is important, because knowing that you found ways to deal with the situation will stand you in good stead for the next time you feel overwhelmed by uncertainty. The overall goal is to move from believing that "uncertainty is always bad and I can't cope" to believing that "sometimes uncertain situations turn out well and, if they don't, I can cope with them."

It can be very useful to record some of your behavioral experiments in your journal, spelling out what your belief about uncertainty is, how you are going to test this, and how it all went, as below.

Your Belief	Your Test	The Outcome
What belief about uncertainty do you want to test? How much do you believe it now? (0–100%)	What can you do to test this belief?	What happened? How much do you believe it now? (0–100%)
If I'm uncertain about where my partner is, I won't be able to concentrate on work. Conviction: 90%	Don't get in touch with my partner for two hours. Measure: Focus on a project.	I spent two hours not getting in touch. I felt very stressed and worried, especially after the first hour. But I did manage to get some work done on the project. Some concentration was possible. Conviction: 80%
If I phone a friend after a long absence, she will be angry with me and won't want to talk. Conviction: 70%	Phone my friend.	She was really pleased to hear from me and we had a long chat. Conviction: 10%
If I don't pack my son's sports bag before a training session, his coach will be annoyed and he will miss training. Conviction: 85%	Allow my son to pack his own bag.	He did forget his lunch and his socks. He spoke to the coach, who found him a spare pair and his friends shared their lunches with him. Conviction: 65%

Learning how to manage worry

Your personal well-being is dynamic, volatile. and requires tending to every day. Remember that you are the steward of your own well-being. Some elements are within your control, others aren't. So, it's important to try to focus on the things that *are*.

Alongside exposing yourself to these small doses of uncertainty, it is also really helpful to reflect on patterns of worrying. Ask yourself, "How much time did I spend worrying today?" Psychologists often use a 100-point scale to get an idea of how often and how intensely somebody is worrying. The scale can be an eye-opener as you may not be aware of just how much you *do* worry. It is also a useful way to keep track of any changes in your worry over time—this is another place to use your journal:

1. Thinking of the previous 24 hours, first choose a term that best describes the extent of your worries: none at all (0), minimal (1–20), some (21–40), average (41–60), a lot (61–80), or extreme (81–100). This will give you a useful anchor point from "no worries at all" to "extreme levels of worry."
2. Then, select the most precise number within that anchor point's range. If you choose "average" in step 1, say, you might give yourself 56. Or if you said you worried "a lot," you might give yourself anything from 61, just above average, up to 80, which is almost into the "extreme" zone.
3. Then consider what you are worrying about, and ask yourself, "Is it possible to solve the problem I am worrying about?"

Some problems are, of course, more amenable to solving. If you are having an ongoing conflict with your partner, making time to talk it through with them might help. But if you are worrying about the possibility of a family member becoming seriously ill in the future, that is something you cannot solve.

If your main worries are amenable to problem-solving, then identify the key elements and figure out what needs to be done to resolve the problem. Try not to get caught up in irrelevant details—this is a common and unhelpful way to avoid confronting the problem. We often need to resolve problems without having all of the information we would like, and so it's important to learn to make decisions with some degree of uncertainty. You need to find a good compromise between gathering excessive amounts of information and avoiding the problem completely.

For worries that are not amenable to problem-solving, confront the worry directly. One surprisingly effective way to do this is to record yourself describing your worry as vividly as possible on your phone. Listen to the recording four or five times, allowing yourself to engage with it and think about it in detail. As you listen, don't try to suppress the worry, just absorb it. This might be upsetting to begin with, but the constant exposure will gradually make the worry less and less threatening.

Do you think worrying makes a difference? It's not uncommon to believe that the act of worrying itself can make a difference. I worked with a CEO of a manufacturing firm who thought that worrying about whether his staff might be cutting corners was crucial to the safety culture of the factory. This became a real burden as he was constantly thinking about what people were doing and several times a day would ask people about their safety procedures. Unsurprisingly, this unsettled his team because people felt his constant checking meant that he did not trust them. I asked him to try worrying about just one staff member for a couple of days, and then another one for the next couple of days, and then assess whether there was any difference in the two employees," performance. He gradually began to realize that his worries made little difference other than making him stressed. He began to get more comfortable with being uncertain and found a more functional coping strategy by having just one session a week with each of his staff to discuss safety procedures.

Worry is often used as an everyday strategy and, as we have seen, it is often not very useful. Many people become very good at managing their worries until there is a crisis; then worry, and what is called "catastrophic thinking," can take center stage.

Dealing with a crisis

Optimal performance often boils down to being able to manage yourself effectively in the unpredictable, uncertain world around you. And there will be times when you will be faced with an unexpected crisis: the diagnosis of a serious illness, the death of a loved one, or the loss of a job. In a crisis, do you try to control everything?

Well, I'm afraid I've got news for you—you can't, and trying to is a recipe for enduring stress.

When you are faced with a crisis or extreme uncertainty, your threat system will be on high alert. There are a series of steps that can be helpful to work through. You may notice your heart beating rapidly, you may feel faint, you may feel short of breath. Deep breathing is powerfully calming. Just take a moment and several deep breaths, remembering to breathe out for longer than you breathe in. Unless there is an obvious escape route, or immediate action is required, don't try to make anything better at this stage.

Once you have calmed yourself and taken the shock on board, observe what's going on in your mind. Are you catastrophizing? Is this a situation that can be resolved in the short term, or is it likely to unfold over a longer period, such as if you are diagnosed with a serious illness?

Try to step back from the situation and look at it objectively, regardless of your feelings and thoughts. This is called "decentering." It's not that easy, but is a really useful technique. Decentering involves shifting your perspective and taking a nonjudgmental view of the situation and how you are reacting. One way to achieve this can be described as the "NOSE technique," and it is a surprisingly effective calming tool. It is something that's worth doing every day as a calming exercise or in any unpleasant situation—you don't have to wait for a crisis.

- Notice what's happening in your body.
- Observe what's going on in your mind and the situation.
- Step back from what's happening in your mind.
- Experience the situation from a shifted perspective: *decenter*.

Expecting the unexpected

Will Greenwood, now retired from international rugby, belonged to England's 2003 World Cup–winning rugby team. We shared a platform at a showcase on "adaptability" to reveal how the England team trained specifically to get used to "dislocated expectations." They borrowed this from the Royal Marine training idea that "whatever you expect is not going to happen." A well-known military technique is to wait until the end of a long and hard training run when the recruits are looking forward to climbing onto a truck to take them to some food and a hot shower. Just as they are beginning to relax, they are told that there's another five miles to go. The rugby team adapted this in their own training sessions and prepared for all sorts of unexpected scenarios: the opposition scoring in the last minute, their best two players being injured, going behind in the opening minutes. The idea was not just to anticipate these specific scenarios, but to embed the principle that things will never play out as you expect.

Bob Bowman, coach to the American swimmer Michael Phelps, was also a great believer in this technique. He would occasionally break Michael's goggles immediately before a training session or a small race so that Phelps would have to swim without being able to see. At the 2008 Olympics, early in the 200m butterfly race, Phelps's goggles came loose and started to fill with water. He swam most of the race without being able to see clearly and the last 50m without being able to see at all—but it did not stop him from winning and breaking yet another world record. Because of Bowman's unconventional coaching techniques, Phelps had mentally prepared for just this situation; he had worked out how to count the number of strokes to get to the end of the pool when he would have to turn, and was unaffected by the inconvenience of the broken goggles.

Chapter Summary

- People differ in how well they can tolerate uncertainty, but we can work on changing our tolerance levels.
- Uncertainty can trick you into making decisions too quickly and encourage you to stick with the familiar.
- Becoming more comfortable with negative feelings and thoughts will help you deal with uncertainty.
- Exposing yourself to small doses of uncertainty in real-life experiences is a powerful way of building your tolerance.

CHAPTER 3

THE FLEXIBILITY OF NATURE

In the 1960s, biologists were fascinated by the nervous system and were on the hunt for the perfect creature they could study to reveal this system's secrets. The nervous system is made up of the brain and spinal cord, and contains complicated tangles of nerve fibers that take in information from the outside world and direct the actions of the animal. Understand the nervous system, the thinking went, and we would gain a deeper understanding of what drives animal and, ultimately, human behavior. The human brain was too complex; the biologists needed a small animal, with a simple nervous system, that could be more easily studied.

The answer eventually arrived in the form of a species of roundworm called *Caenorhabditis elegans*, and the biologist Sydney Brenner began his now-famous "worm project" in the MRC Laboratory of Molecular Biology at Cambridge University in 1963. These worms are still studied in laboratories all over the world and have since become the best-understood creatures on the planet. They are also the foundation of our understanding of how brains work.

Agility is built in to how even very simple brains work

Brenner and his team discovered many fascinating things about the brain from *C. elegans*, including the fact that agility is hardwired into even these simple worms. *C. elegans* has precisely 302 brain cells—or neurons—with a total of around 8,000 connections, or synapses, between them. This simple nervous system results in quite an inflexible range of behaviors that occur under fairly constrained circumstances, meaning that any action the worm takes is usually clearly connected to a specific cue. For example, the cue of feeling "cold" is tightly linked to the action "move away," and a drop in the level of oxygen in the soil triggers the very specific response "escape."

But there is a surprise. What has been discovered in much more recent research is that in spite of this rigid cue-action connection, the worms can still be very flexible in their responses. When a threat is detected, the worm will begin an automatic escape routine, but the *type* of escape is very variable. The scientists realized that a small number of the worm's neurons—called "command interneurons"—are always active and can cause spontaneous turns or reversals of direction in a seemingly random fashion. The worm behaves in unexpected ways, turning left or right, for instance, even when there is no external signal or trigger to do so. This spontaneity allows a worm to learn from her experience; for instance, she might find an unexpected food source by turning left. So, even the simple *C. elegans* nervous system has agility built in.

Agility in the natural world

From single cells to the most complex of biological systems, agility and flexibility are fundamental. Almost all fish, for instance, can switch sex at the embryonic stage to benefit their species. If their population begins to decline, perhaps because of chemical pollution in the water or major changes in temperature, embryonic males will transition into females to help ensure that the species survives. The opposite also happens. Take the sea bream as an example, a species that is heavily fished in the cold waters of the North Atlantic. Because larger fish, which are usually male, are favored by trawlers, some females will change sex so that the balance is preserved. Likewise, when a school of reef fish loses its single male, the largest female begins acting like a male within hours and will produce sperm within ten days. This remarkable feat is achieved by means of hormonal changes that begin the transformation of their organs from those of one sex to the other. This adaptive ability to switch sex explains the high degree of diversity that we see in fish—there are some thirty-three thousand different species of fish as compared to just six thousand species of mammals.

Major natural adaptations also occur in the behavior of mammals to allow individuals and groups to deal quickly with threats. These changes can take several generations to become well established, and different species can deal with common threats in quite different ways. For example, extreme cold and lack of food in the winter might cause some species to migrate to warmer climates where food is more plentiful, while others might hibernate throughout the cold season; their body temperature drops so they don't need to eat and so don't have to venture out into the cold. While these are completely different approaches to the same problem, both are highly effective.

At a more fundamental level are bacteria, which can adapt by stealing genes from elsewhere, whether from other cells, including other bacteria, or even from DNA molecules that are floating in the environment. This system of "horizontal gene transfers" allows the bacteria to "buy in" new techniques and habits that help them to thrive in many different environments. It is this adaptability that helps bacteria to develop a strong resistance to antibiotics. Viruses are similarly adaptable. They can mutate rapidly to try to find a fit with a new host. We saw exactly this with coronavirus. As populations were vaccinated, the virus was constantly changing itself. This was biological evolution, and agility, in action, however much we might not like the outcome.

Luckily our immune system also has a range of options available to fight back. Gerald Edelman, the American biologist and winner of the Nobel Prize in 1972, discovered that the human immune system works by producing millions of antibodies, each with a slightly different shape, so that at least some of these antibodies are likely to fit, and hence be able to block, the chemical receptors of an invader. This gives the immune system a remarkable degree of adaptability. By producing a defense for every possible scenario, the immune system is ready to repel nearly every type of attack.

What we see again and again is that many genes, immune-system cells, and biological structures that are quite different from each other can often perform the same function by different means. This tendency of biological systems to do the same thing, even though they are structurally quite different, is called "degeneracy" in biology and it is why the system is so agile and resilient. A good example is where several quite distinct processes in the body can convert food to energy very rapidly. This means that metabolism (converting food to energy) is a very robust process, so even if one pathway cannot function—due to disease, for instance—another pathway takes over so that the entire system continues to work well.

Flexibility is also central to how our brain works

Diversity is also typical of our own nervous system. Compared to the 302 neurons and 8,000 synapses in the C. elegans brain, the human brain is estimated to contain 86 billion neurons and hundreds of billions of synapses. This means that we have a much, much greater capacity for agility. The downside is that this flexibility comes at a cost; while our brain takes up only around 2 percent of our body weight, it consumes a massive 25 percent of our energy. However, it is also incredibly sophisticated, and allows us to manage multiple goals at the same time.

This works well because the brain communicates with the body, and within itself by means of complex patterns of connections among neurons that flow like waves across the brain. No two neurons are identical in shape and size, and each neuron typically receives messages from thousands of other neurons. What this means is that in just one tiny area of brain tissue there are billions of connections, or synapses. These astonishingly intricate patterns of connectivity are not only unique to each of us but are also not fixed and can change over time. This complexity indicates that these connections are highly unlikely to be preprogrammed—instead, agility is necessary so that the system can react rapidly to changing circumstances.

This works in a straightforward way. When we learn something new, certain neural connections strengthen. This is referred to as the Hebbian rule, in honor of the Canadian psychologist Donald Olding Hebb, who discovered that "cells that fire together wire together." In other words, when specific neurons are activated at the same time the connections between them intensify, and they all react as one as if they were a physically connected circuit. But they are not physically linked; they are simply coupled together by their role and the fact that they are activated at the same time. If a period of time goes by and these circuits are not used regularly, then the strength of the alliance will gradually decline, and the "circuit" will fade away.

These circuits of neural connections emerge and transition in response to a changing environment and allow our brain to respond flexibly to almost any scenario. To add to the complexity, many different patterns of neural connections can result in the same outcome—another example of degeneracy. When we are learning to emit a new spoken word, for instance, a circuit of connections between neurons that combine a thought in our head with very specific muscle movements of our tongue, mouth, and vocal cords becomes gradually strengthened as we learn how to pronounce that word in the right context. If we have just been to the dentist and had an anesthetic injection, however, the usual neurons that activate muscles in the tongue may no longer be working properly, so another circuit rapidly springs into action, using slightly different muscles to perform the same function. This explains why you might talk a bit strangely until the anesthetic wears off. This happens all the time. If a neural circuit is disrupted or distracted, another circuit can rapidly take over.

The newly discovered importance of the gut

Beyond internal communications in the brain, we now know that the connection between our brain and our body, especially our gut, plays a far bigger role in determining our thoughts and behavior than we previously realized. It is the brain's number-one priority to maintain a balanced bodily state by coordinating all of the information that comes to us from the outside world with the system that controls how we act and the workings of our internal physiology. This three-way interchange—the gut-brain-surroundings axis—ensures that we have sufficient metabolic resources to survive and thrive, and this is what our brain really cares about. This is a reflection of switch craft in operation at a biological level; being alert to changes in the environment, being aware of our internal state, and being agile are hardwired into the way our body and brain work together.

Neuroscientist Lisa Feldman Barrett describes this as an internal "body budget" that your body and brain is constantly keeping track of. Your body has limited resources and therefore, any time your brain prepares to engage in an activity, whether thinking or moving or phoning a friend, it is calculating whether this is a worthwhile investment. Is there enough in the bank to justify the expenditure? If you are lacking in a particular nutrient, for instance, your brain might suppress all other metabolic processes and actions to prioritize finding that nutrient.

This constant budgeting and anticipation means that your brain is constantly making predictions about the best course of action to take, as we have mentioned in earlier chapters. You may have to stop work and go for something to eat, you may have to sleep, you may need to exercise, you may need to find a particular food. Your brain is built to be agile and alert to your surroundings. From moment to moment your brain is predicting what's likely to happen next and then rapidly feeding information back about what happened. If things happened as expected, the predictions carry on as normal. If the outcome was not as expected, there is an interrupt, an error signal, and this is stored in our brain networks for future reference. The implications of this system are profound. Instead of being a passive observer of reality, your experience of the world is constructed within your brain. Your brain makes a best guess, and then cross-checks this guess against incoming information.

Think about the tennis player Roger Federer waiting to receive a serve. Top professionals like Novak Djokovic regularly serve at over 200km per hour. This gives Federer less than half a second to respond and this is not enough time to move to the right position, prepare his racket, and make the shot. Instead, what happens is that Federer's brain makes an unconscious *prediction* about the likely location of the ball, plans the necessary action before Djokovic hits the ball, and then rapidly readjusts if something unexpected happens. Eye-tracking studies of tennis players show that average players look at the ball during a serve, while elite players look at their opponent's arms, hips, and general body movements to predict how they should position themselves even before the serve is hit. This is how our brain works, by anticipating rather than reacting, and this is just as true in everyday life as it is on the tennis court.

And the predictive nature of the brain is why failures and setbacks are just as important for learning, if not more so, than are successes.

Chapter Summary

- Agility is hardwired into our brains. Even a simple worm with just 302 brain cells is agile and can learn from experience.
- Agility is essential for survival and is endemic across the natural world, from fish changing sex, to how the immune systems of mammals function, to how bacteria and viruses survive.
- The major components of switch craft—being alert to your surroundings, being aware of your internal state, and agility—are hardwired into the way our brain and body work.
- Flexibility is crucial for the predictive way that our brains work.

CHAPTER 4

AGILITY AND RESILIENCE

We can all point to people who seem to cope particularly well with whatever life throws at them. So, what is it that sets these resilient folks apart? Is it just part of their personality? Is it preprogrammed in their genes? There are no clear-cut answers, yet. Resilience has become a buzzword in recent years, with a frenzied focus of scientific research, but there is still a great deal of uncertainty in this field, and ongoing debate on what we even mean by "resilience." In fact, recently it's become clear that researchers may have been looking in the wrong direction, and asking the wrong questions.

What is resilience?

Being resilient means doing better than expected

The way in which we understand resilience has changed considerably over the past decades. There was a time when a resilient person was considered to be someone who was unaffected by stress, or someone who was thriving and happy all the time, or who could "bounce back" to normal after any adversity. But of course, there is no such person. Stress affects all of us to some extent. We now know that resilience is an ongoing and dynamic process of adapting well in the face of life-changing situations and serious stress; sometimes we do well, other times not so well. The best way to think about a resilient person is as someone who is doing "better than expected" in light of what they have been through. You may be struggling and feeling anxious and depressed after a trauma. But if that's "better than expected" given what you have endured, then you are still resilient.

We are way more resilient than we think

When we look at what happens to people following a major trauma, we find that there are many different pathways to resilience and success. What's often not appreciated is that we are naturally highly resilient. Study after study shows that most of us, two-thirds in fact, come through even major trauma in a resilient way. Even when facing life-changing events, such as earthquakes, the death of a loved one, terrorist attacks, losing everything and becoming a refugee, two-thirds of people adapt well and are able to function effectively afterward.

Resilience is not just in the mind

Resilience has as much to do with what we *do* and what we *have* as how we think and feel. Rather than searching for the magic juice that makes one person more resilient than another, it's more realistic to look at resilience in a holistic way. In fact, recent research tells us that there are a variety of "protective" factors—practical advantages that impact resilience—that enhance our resilience and, importantly, these influences can be modified. For instance, there is growing evidence that our relationships and social support are critical, and that resilience often boils down to being able to negotiate to get what you need in a difficult situation. This might include actions such as seeking help after things go wrong rather than withdrawing from the world, reducing use of alcohol and drugs, choosing healthy behaviors such as a better diet and exercise, as well as resisting unhelpful habits like excessive worrying and rumination. The availability of external resources, such as whether we're lucky enough to have access to basic services, is also of course crucial.

Leading resilience researcher Michael Ungar uses the fairy tale of *Cinderella* to illustrate this. "We misunderstand the story," he says. "Everyone thinks that Cinderella's transformation is due to her inner characteristics, her beauty, her kindness, her optimism, and her grittiness." While these are important, he suggests what was really vital was her fairy godmother. "Think about it," he says. "The fairy godmother provided her with what she needed to go to the ball; without fine clothes and a carriage to take her there, she would never have been able to meet her prince."

If your house has been flooded, what's helpful is good insurance to rebuild; to recover from a serious illness, what you need is good access to medical treatment and time to recover; if you have been dispossessed and become a refugee, a package of financial and social support around you is likely to have the most impact. Several studies of resilience among refugees resettling in other countries show that an important predictor is the presence of a least one child who can speak the language of the host country. Being able to communicate in a new and strange country gives you a better chance of getting what you need. We might not all have the same physical resources but the ability to negotiate for them can be transformative.

There is no one-size-fits-all version of resilience

These "protective factors" are not necessarily within our control and will be different in different situations. Take the impact that a teacher can have on a young child. Lots of research tells us that an encouraging and supportive teacher is often a vital turning point in a child's life—and can shift them from a path to disaster to a trajectory to success. But other studies have found that for many students a good teacher makes little difference to a child's resilience. How can this be? The answer is very straightforward. If a child comes from a highly supportive family who provides them with a secure base and lots of encouragement, then the impact of a school, or a specific teacher, is relatively small in comparison. In contrast, if a child is coming from a very difficult family background where there is little support and encouragement, then the impact of a supportive teacher can be immense. So certain factors—a supportive teacher in this case—are not "protective" in and of themselves. Instead, the degree of protection depends as much on the circumstances of the child as on the behavior of the teacher.

What this means is that the quest to find the essence of resilience within an individual is doomed to failure because it is so different depending on the context. The usefulness of our various personal abilities and habits is determined entirely by the nature of the crisis that we are trying to deal with. What might help someone cope with losing their home and all their possessions in a house fire may not be the same things that help someone trying to save a struggling business. This is also true in more everyday situations. For instance, working out regularly in the gym and keeping physically active can be a great way to boost resilience, but if you are recovering from a serious illness, pushing yourself to exercise might be counterproductive and undermine your resilience. The latest view of resilience is a more subtle one, and tells us that when faced with adversity, each of us needs to navigate our own unique course through these difficulties.

Agility is essential for resilience

Because resilience is a dynamic process with many facets rather than an inner trait, different solutions are required for different situations. This is why mental agility is so vital; it helps us to negotiate and find the resources that we need. Sometimes these are internal and sometimes they are external. If you are being harassed and bullied by a belligerent boss, spending an hour every morning meditating to help you deal with stress may not be the most helpful strategy. You might be better off trying to find a new boss or finding ways to get away from that situation. If you are trying to cope with the stress of long-term chemotherapy, on the other hand, meditation might be exactly what you need.

Studies by my lab group have explored resilience to anxiety and depression across the critical teenage years. Our approach is to first get a clear idea of what people have been through so that we can assess whether they are doing better or worse than we might expect. The type of things we are interested in range from major events like death of a parent, family violence, and divorce to more positive situations like moving house. We find that being from a more affluent family, with access to more resources, undoubtedly gives you a resilience advantage, as does being male or having a healthy degree of self-esteem. The reasons why are clearly complex, particularly in relation to gender. In studies of corporate environments, women easily match their male counterparts in terms of self-esteem and ability to negotiate. However, there is also evidence that some of the related traits—the ability to ask for what you need, for instance—are praised in men but seen as socially undesirable in women.

How can we become more resilient?

We can cope with problems by either focusing on the problem or focusing on how we feel. Our findings on gender and resilience also reflect the fact that women typically experience more anxiety and depression than men. We don't really know why this is, but one clue might come from the different ways that men and women often seem to respond to stressful situations. When faced with stress, men will typically try to solve the problems, while women are more likely to ruminate about them. These coping strategies aim to either eliminate the problem—"problem-focused"—or to work with the source of stress in some way—"emotion-focused." And there is a mountain of research on the costs and benefits of each.

Emotion-focused coping relates to the ways you attempt to manage stress. You might try things like praying for guidance and strength, distracting yourself, talking to friends about the problem, eating comfort food, taking drugs, drinking alcohol, or trying to think about the situation in a way that reduces its emotional impact. Some of these approaches are useful ways of coping when you have little control over the source of the stress. However, when there is a possibility of controlling the root cause of the issue, problem-focused strategies are generally a more effective solution. This involves evaluating the pros and cons of different solutions to try to directly confront the stressful situation. Imagine a scenario in which it's clear that you are being discriminated against or looked over in your workplace when promotions come along. If you opt for an emotion-focused approach, you might complain to your friends outside work, try to meditate, relax, and not think about it. All these approaches might help you to cope in the short term, and might help you to feel better, but they will not resolve the issue. A problem-focused approach might involve raising the issue with your line manager, starting a campaign with a union, or simply looking for a new job.

Ask yourself whether you are flexible in your approach

As we saw in Chapter 2, it is always important to step back and consider whether you realistically have any control over a demanding encounter. Only then can you decide on the best strategy. Is the problem one that you can deal with directly? Or is it one of those situations that you must just roll with? If you need to have a wisdom tooth out, you just have to deal with it. Some people are inflexible in that they overvalue control and try desperately to manage every stressful event. Others rigidly consider all stressful encounters to be uncontrollable. Those who are more agile will judge some situations to be controllable and others to be uncontrollable. There is a growing body of research telling us that taking a flexible approach to coping is linked with greater happiness and well-being. Think about yourself. Do you tend to perceive most stressful events as controllable, or uncontrollable? Could you be more flexible in your approach?

When I talk to young people at school events, flexibility is a constant theme that emerges. When explaining how they cope with difficulties and challenges, young people who cope well usually describe how they regularly try out lots of potential solutions. Those children who develop a variety of different ways to deal with their difficulties, and who are agile in how they implement these different approaches, are most likely to overcome the adversities they face. In fact, an intriguing conclusion has emerged from these discussions showing that adversity can sometimes be a good thing. In an echo of the old mantra, "What doesn't kill you makes you stronger," those children who have had to deal with many problems often develop a range of ways to cope that they can use flexibly and effectively, and this boosts their resilience. This is backed up by our own studies that look under the hood and tell us that more flexible thinking styles are also associated with a greater degree of resilience.

Growing up under severe deprivation can of course cause serious problems for a child's development, general health, and ability to learn and thrive. For this reason, social scientists and policy makers often focus exclusively on what is called the "deficit model." This means looking at what's *wrong* with kids from poor and deprived backgrounds rather than what's *right* with them. But the deficit model misses interesting information about how children like this learn to adapt to their immediate environment. Children living under very tough conditions often fine-tune their mental processes to deal as best they can with the challenges.

Hardship and difficulties can boost your resilience

During one of my studies I met a teenage boy called Andy. Both Andy's parents were drug addicts, and his mother was often violent. Because of this, Andy had been taken in and out of care regularly since he was about eight years old. "There came a point in the day," Andy told me, "when I would notice the signs that she was beginning to turn." What kind of signs? I asked. "Her voice would change," he explained. "She would stare at me hard." At that point, Andy had worked out he should disappear. He became a master at noticing even the subtlest of changes. As long as he stayed out of the way he was safe, but if he hung around the kitchen when she was in that kind of mood, he knew there would be trouble.

Being able to notice the slightest hint of anger was vital for a child, like Andy, living with a violent parent. Andy had become highly vigilant for threat, which could create difficulties in the wrong circumstances, but he was also very resilient and coped well with many difficulties. It was no coincidence that he was highly socially intelligent and helped many of his peers to sort out their problems. His teachers reported that he was becoming a natural leader. What Andy had learned was to fine-tune the way he behaved to the specific demands of the immediate situation.

This ability to match our actions and our thoughts to the situation at hand is vital for a resilient outcome. If Andy had maintained the same degree of vigilance that he needed at home when he was in a happy relationship later in life, he would almost certainly not be thriving so well. This is because it is the *flexibility* with which these processes can be modified, and adapted, across a range of different situations that really predicts who thrives and who struggles with various life challenges.

The message for all of us is that agility is a vital skill to help us live a more resilient life. We cannot avoid difficulties and challenges, therefore the more ways we have to cope with them, along with a flexible approach to change tactics if needed, then the more resilient we are likely to be.

Being agile is at the heart of resilience

The more we try out a variety of solutions to solve different life problems, the more deeply we embed a profound knowledge of the type of strategy that might work in specific situations. As we move through the book, you will see that resisting the rigidity that is so alluring when we have found a comfortable niche is at the heart of thriving. And if you can keep your behavior and feelings agile, you will develop a suppleness of mind that allows you to adapt and optimize your performance, no matter what life throws at you.

Let's take another example. In the early 1980s Jason Everman blew up a toilet at his junior high school with an illegal firecracker. He was given a couple of weeks," suspension and sent to a psychiatrist. Jason refused to speak a word during the sessions, but as it happened the psychiatrist was a guitar buff and started to play some tunes and teach Jason a few chords. It became a family joke that these were the most expensive guitar lessons ever. The psychiatrist hoped that playing the guitar would open Jason up. It didn't: but it did transform Jason's life. Just a few years later a childhood friend invited him to join a newly formed band that was looking for a bass player. They called themselves Nirvana. Jason, who still suffered from dark moods, was fired from the band just a few months before they got their big break.

Jason adapted quickly, however, and a short time later was asked to join another band that was even bigger than Nirvana at the time, Soundgarden. This was a band he had always wanted to be part of and he was overjoyed. The following year was a heady mix of touring throughout Europe and the United States, honing their craft, and planning a new album, which Jason used the last of his savings to help fund. When they returned home, on the verge of stardom, Jason was kicked out of the band—again. The following year, Soundgarden's new album, which Jason had helped to finance, went double-platinum and Nirvana, of course, became the biggest rock band on the planet. In an interview with the *New York Times* many years later, Jason admitted that this had been a devastating blow and it took him months to recover.

He moved to New York and joined yet another band, Mind Funk, but soon realized that he didn't want to be the guy in his fifteenth band reminiscing about his heady days in Nirvana and Soundgarden. "I was in the cool bands," he said, "but I was psyched to do the most uncool thing you could possibly do." Out came the nose ring, off came his long hair, and at the age of twenty-six Jason joined the army. He excelled at that too, ending up serving with US Special Forces in Afghanistan and Iraq and being awarded numerous medals for bravery. The dark moods did not go, a military colleague told the *New York Times*, they just didn't matter anymore: they didn't interfere with his work. When he left the army, Jason went to university and received a bachelor's degree in philosophy at the age of forty-five from Columbia University in New York, followed by a degree in military history from Norwich University. He is now learning to sail and aims to sail around the world alone because it is "the classic man versus nature conflict." He intends to "keep engaged with the world, with life" because "getting old is an adventure in itself."

Pressure and stress can magnify the allure of the familiar. As we saw back in Chapter 2, when we are unsure of ourselves, or simply frightened, our natural tendency is to revert to what has worked before rather than keeping an open mind. Jason could have stayed playing in bands—it was familiar to him and he was good at it—but he also realized it just wasn't working for him. So, he made the leap and moved to a completely new way of life. His agility allowed him to switch rather than sticking with the familiar.

To enhance our resilience, we need switch craft

This outlook of seeing life as an adventure, keeping an open mind, and maintaining the agility to try out different things shows switch craft, and resilience, in all its glory. Our ancestors thrived because they had the capacity to adapt to a wide range of environments; similarly, people who are resilient are able to use a variety of strategies that allow them to adjust to whatever life throws at them.

We can all improve our capacity for resilience, and we can do this by developing the four pillars of switch craft that we will work through in the following chapters. Being agile means that we make use of whatever traits, abilities, or resources that best fit the situation. To build our resilience we need to open our minds, assess the situation with a clear-eyed focus, and decide what is the best course of action. The more we master the principles of switch craft, the more resilient we are likely to be.

Chapter Summary
- Resilience is not about being unaffected by change and adversity, but doing "better than expected" relative to what you have been through.
- Resilience is not a magic potion, it is dynamic and influenced by many different things, which is why it is closely linked to agility.
- Resilience is determined as much by what we *do* and *have* as by what we think or feel.
- Overcoming challenges and experiencing adversity often helps boost our resilience.
- Agility, or flexibility, in coping with stressful situations is the key to building resilience.

THE FIRST PILLAR OF SWITCH CRAFT

Mental Agility

CHAPTER 5

THE BENEFITS OF MENTAL AGILITY

It was a bright sunny morning and Paddi Lund was on the verge of a mental breakdown.

On the surface, Paddi was doing well. He was financially successful with a thriving dental practice in a bustling suburb of Brisbane in Australia. But he had become obsessed with trying to build a business empire rather than doing the dentistry job he loved. He was constantly stressed, and his personal relationships had deteriorated. He was profoundly unhappy and was getting no pleasure from life.

Paddi realized that he had to do something radical. After ten years of building a successful business, he asked himself what he actually enjoyed and how he could maximize those experiences. He realized part of the problem was that he only enjoyed the company of some of his patients—the majority of them he found difficult. So he decided to downsize and work with only those patients whom he liked. This meant getting rid of almost 80 percent of his client list. He removed his practice from the telephone directory, deleted his website, took all signposts down from his buildings, and asked his remaining patients to recommend him to their friends. He figured there was a better chance that he would get on well with the friends of people he already liked. Paddi then transformed half his building into a café, with the aim of creating a happiness-centered dental business. A few years later, he was working about twenty-two hours a week and had more than doubled his income. His reduced workload meant he was spending lots of time with his family and friends and taking up new hobbies. Most importantly, Paddi was happy.

What is mental agility?

All of us will face differing degrees of stress at different points in our lives. "Into each life some rain must fall," mused the poet Henry Wadsworth Longfellow. The dentist Paddi Lund found his own innovative and agile way to deal with a stressful life and made a major change. But there are many ways to manage stress, and not all of them are radical. So it's important to have a range of different ways to manage stress and anxiety. This is part of mental agility—having a flexible approach to a complicated world. No one approach is likely to be effective for every situation. As the American psychologist Abraham Maslow warned, "I suppose it is tempting, if the only tool you have is a hammer, to treat everything as if it were a nail." What Maslow meant was that we tend to tackle problems with the tools that we have most readily available rather than looking to see whether there might be a better way.

The same is true of mental processes. If you habitually deal with a problem in a certain way, then that might work well some of the time, but at other times it might be exactly the wrong thing to do. We have seen that turning inward and rigidly looping a problem around in your head—worrying—is a tried and trusted method when you are feeling overwhelmed or uncertain. On some occasions this can help, but most of the time it only makes things worse. There comes a point when you need to get out of your head, away from the internal chatter, and as Paddi Lund discovered, think agilely to find solutions outside yourself.

How do you know if you are mentally agile?

Several years ago, I realized that the available measures of mental agility were inadequate. So, along with my lab team, we developed a new questionnaire that included the main components of psychological flexibility and spent almost two years posing those questions to hundreds of volunteers. You can fill out the resulting *Mental Agility Questionnaire* below to see how mentally agile you are in comparison with the thousands of students, businesspeople, and athletes we have tested so far.

Mental Agility Questionnaire
Please rate the degree to which you agree or disagree with each of the statements on the scale below. Think carefully about each statement and answer honestly.

6. Strongly agree
5. Agree
4. Slightly agree
3. Slightly disagree
2. Disagree
1. Strongly disagree

1. I am optimistic about the future.
2. I am much more open to change than my friends.
3. I am good at getting used to different situations.
4. I sometimes do unusual things.
5. I am confident that I can adapt to new situations.
6. I know that things will always change—it's just the way life is.
7. Once I start something, I find it fairly easy to stop if I have to.
8. When I encounter difficulties, I try lots of different ways to find a solution.
9. I am good at switching quickly from one thought to another.

10. I am good at coping with the unexpected.
11. I find it exciting when things are changing rather than getting stressed.
12. I sometimes think in ways that are very different from other people.
13. I am keen to learn from other people.
14. I am good at juggling different ideas at the same time.
15. I find it easy to balance my long-term goals with what I want to do in the short term.
16. I have learned that some things I do work well in some situations but not in others.
17. Most things in life are not black-and-white—they are much more complicated.
18. I am very good at noticing when people's mood changes.
19. I am able to learn from my mistakes.
20. I am good at changing my mind when I can see that it works for others.

Now add your scores together to get a number between 20 and 120. Put simply, we can work on the following assumptions:

1. A total of between 20 and 60 indicates "Low Agility" or "Inflexible".
2. A total of 61 to 79 is on the lower end of "Typical Agility."
3. A total of 80 to 99 is on the upper end of "Typical Agility."
4. A total of 100 to 120 is "High Agility" or "Flexible."

How did you score? Whatever your result is, remember that this is not a fixed state. You may be fairly inflexible now but after working through the exercises in this book you should find that your mental agility increases. If you are already agile, that's great, but don't get complacent—it's always something that you can continue to improve.

The power of an agile mindset

Thanks to his agile mindset, Paddi was able to make changes that transformed his life. From the depths of his despair, his mental openness to act on his gut feelings gave him the freedom to choose an optimal way of living. Just think what such an agile mindset could do in your own life. This is the first pillar of switch craft we will be tackling—how to build your agility.

Escape from the limits of a constricting mindset

Agility sets your mind free to consider alternative possibilities and allows you to escape old-fashioned mindsets—what some psychologists have called the "tyranny of automaticity." This "tyranny" refers to the difficulty in breaking away from our habitual ways of doing things. The mantra "We've always done it that way" is just not a good enough reason to do anything. The status quo may be comfortable, but it's important to question things: Are our past habits and ways of doing things really serving us well?

Stick or switch?

Being agile spans a range of human abilities but often boils down to a simple decision: Do you stick or switch? Switching is draining and therefore should only be done if it's necessary and useful. The energy-sapping nature of switching is also one reason why we have a natural tendency to stick, putting us at risk of mental arthritis.

Think of a time when you were faced with a real difficulty, a feeling of unhappiness, or just a general sense of unease. You might well have found yourself asking whether it was time for a change. This is not a simple decision. You don't want to quit just because something feels hard. No one would win Olympic medals or succeed in any walk of life if they gave up when real effort, or "grittiness," was required. But there are times when relentless grit is not enhancing your life and sometimes amounts to pushing against a brick wall where there is very little chance of success. A reluctance to change direction might cultivate unhappiness and close off other options. Making plans and sticking to them relentlessly, regardless of the outcomes along the way, is just not the way life works.

Can you think of a time in your life that you stuck with something, or someone, for too long? Were there signs that a change was needed? What held you back from changing? Thinking about this might help you to start assessing whether you tend to stick with things when it's obvious that it's time to change. Or you may be the opposite, you may tend to quit well before you should. This fundamental decision—stick or switch—applies to the big things in life as well as in the small everyday decisions that we make on a regular basis.

Both strategies are important for different goals and moments in your life. When you are working on a single task that requires repetitive practice, like training for a competition or exam, then persistence is required. When you are working on a complex project with many different elements, however, you may need to switch regularly between different aspects of the task.

If the world stayed the same, grit would be enough but, as we've seen in recent times, the only thing that is predictable in life is unpredictability. People don't always do what you expect of them, surprises happen, and new inventions leave us with skills that are no longer needed.

A brain that can switch more easily will help us thrive

Recent and cutting-edge science has shown that what psychology researchers tend to call "psychological flexibility"—what I call "mental agility"—is directly linked to happiness and success. Whatever you call it, it has been measured in many ways, usually by questionnaires and interviews. My own studies have tried to look "under the hood" to try to see whether we can figure out how the workings of the brain can facilitate our mental agility in our actions more broadly. This work explores the fluidly of how our brain can stick or switch. We will explore this work in more detail in the next chapter, but for now the closeness of the links we have found between a range of different biases may provide a clue. Let me explain by looking at one of my own studies with teenagers in more detail.

The first step in our study was to measure three different types of cognitive biases in the teenagers by directly assessing if they direct their *attention* toward unpleasant or pleasant images, what they actually *remember*, and how they *interpret* ambiguous scenarios that we pose to them. This gives us three separate measures of biases: one for the attention system, one for the memory system, and one for the interpretation system. As an example, to measure a bias in memory we present people with several words, some unpleasant (cancer, failure) and others pleasant (party, success), and later ask people to recall as many words as they can. People who are prone to anxiety and depression tend to selectively remember far more unpleasant than pleasant words while those who are happier tend not to show this preference for the negative. This fits with lots of studies showing that people who are thriving tend not to remember much of the negative stuff that has happened to them.

After collecting this data, we then looked at how these different types of biases were linked with each other. What we found was fascinating: all three biases toward threat and negativity were much more tightly connected with each other in teenagers who were struggling with anxiety and depression. So for example, when a bias in one area, a traumatic memory, say, is activated, it quickly leads to biases in attention as well as in how ambiguity is interpreted. Think of it like an old-fashioned telephone exchange—or network—in which different extensions (biases) are connected to each other so that if you call one of them all of them will ring. What this means is that, for some teens, all three negative biases become activated once any one of them has been activated. This means that ultimately their brains are triggered to pile negative thought on top of negative thought, leading to further increases in stress and anxiety.

The connections for those teens who were thriving were quite different. Here the links between biases were much looser. It's a bit like a disconnected network where the links between certain extensions (biases) are disconnected. So, if a bias in memory becomes activated, that does not necessarily activate a bias in attention or interpretation. This means that some teens are more able to think about negative things without setting off a range of negativity and associated biases across the system.

We still don't fully understand why this happens. But what is clear is that a looser and more agile system leaves people in a more open frame of mind, with a capacity to consider many possibilities, including actions that may not have worked out well in the past. This loosely connected setup means that people are less likely to get stuck in their ways, making it more likely they will thrive. And the good news is that we can train our brains to be this way and become more agile.

How can we help our brain to become more flexible?

This hidden feature deep in the brain is ultimately reflected in an ability to shift a course of thought or action in response to changing demands. One way to foster these more flexible connections in your own brain is by making tiny adjustments in how you interpret the events around you. One simple technique is to start paying attention to some things that annoy or upset you, and see whether you can find a different way to interpret them rather than allowing your mind to go to the most obvious, negative shortcut.

Let's say you are upset because your friend rarely contacts you—

it is always you who gets in touch. You may feel she is not that interested in seeing you. But perhaps there are some other equally plausible explanations? Perhaps she is so busy with work or with her family that she just never gets around to it? Perhaps she thinks that you are busy and would not be interested in seeing her and so she waits for you to suggest getting together? When you find yourself in this kind of situation, actively try to think of several other interpretations that aren't negative reflections on you.

The idea is to constantly challenge your well-worn interpretations. Done on a regular basis, this exercise will help you to loosen the connections in your mind so that they become more open and flexible and allow you to entertain different possibilities. This will allow you to adjust and tweak how you view things and will gradually undo the rigidity of your mind, ultimately changing the way you act.

Agility and success

Despite the effort involved, becoming more agile will help you to go with the flow. A helpful side effect of increasing your agility in how you think is that it will help you to become better at deciding when to stick and when to switch. We can see this most clearly in sport, where watching an elite athlete at the top of their game shows us how tiny adjustments underpin mental and physical agility. This capacity only comes with practice, practice, and more practice.

Take the Northern Ireland soccer player George Best, who is widely regarded as one of the most naturally gifted soccer players to have ever lived, even compared with superstars like Lionel Messi or Diego Maradona. Watching him play was often described as watching poetry in motion. Even on bumpy pitches, he seemed to always get a good bounce. Whether the ball spun left or right, it invariably arched directly onto his foot. With fluidity and grace, the ball and

Best seemed to move together in a flawless dance. Of course, the bounces didn't actually go with him any more than they did for any other player. He had the physical intelligence and ability to notice the angle of bounce and adjust the arc of his own body to make sure that he was in the perfect position. This was mental and physical agility combined in seemingly magical unison.

We may not reach the sporting heights of people like George Best, but by practicing our mental agility we can improve our ability to choose the right strategy for the right moment. The more effectively we can do this, the more seamless our life will be. Just as an elite athlete can make it look easy, those who are mentally agile can make it look like they are sailing through life with few problems. They are not; it simply looks that way because they are making minor adjustments and adaptations all the time. And these adjustments are necessary; agile people do not change for the sake of it, they change when change is needed.

Mental agility is not about changing for the sake of it, but rather, changing in harmony with your circumstances. This is important to keep in mind. Being agile means switching—or sticking—to adopt the best approach for the task at hand. Keeping young children amused, managing a complex project, negotiating a business deal, conducting a long-term relationship: all require mental agility. In the early days of my academic career I remember the terror of facing over five hundred students in a large lecture hall and trying to hold everyone's attention for the full hour. I quickly learned that the trick was to watch my audience closely. The temptation, especially when I was nervous, was to keep going doggedly from my notes until all of the material was covered. But I realized that if I lost people in the beginning there was little point in continuing. Once I became a bit more experienced, and less anxious, I would look out at the sea of faces more frequently, to gauge whether they were still listening or understanding. Sometimes, following the introduction of a difficult concept, I would see confused or blank expressions. I learned to then adjust, to explain the same thing in a different way, to try to find a better analogy, perhaps several times, before moving on. Being responsive to the students and staying agile was vital, even at the expense of not covering all the material that I had hoped to in the lecture.

Agility also helps to achieve success in business

The toy manufacturer LEGO is a good case in point. In the late 1990s and early 2000s, LEGO was in real trouble. Sales of the company's brightly colored plastic bricks, loved by children the world over, had been in decline year on year. Consultants rushed to the remote Danish town of Billund, where the toymaker has its HQ. They advised that LEGO must innovate and develop a new line of toys. LEGO pumped out creative idea after creative idea over the following years, but sales continued to slump while debts rose. While many of these new toys were innovative and fun, they just didn't engage LEGO's core audience: children who liked to make things.

In 2004, LEGO appointed a new CEO, Jørgen Vig Knudstorp. Knudstorp realized that LEGO had forgotten its core product: the plastic brick. What was needed was innovation *around* the box. What could be done, he wondered, to encourage more children to engage with the humble brick?

Looking closely at their core audience, it was obvious that LEGO was dealing with a generation of children who had grown up with technology. In an agile move, Knudstorp began to look at digital technologies that were complementary to the basic LEGO brick, rather than looking for completely new toys. As a result, LEGO created a hugely successful line of robots, made in the real world with LEGO bricks, that could be programmed to move in different directions by using an app. This innovation helped to bridge the gap between virtual play and real play. A child could now construct a LEGO house within a virtual digital world as well as in the traditional way in the real world. This new digital element attracted adults too, resulting in even higher sales figures. Their success has led to LEGO being called the Apple of toys, with sales in the US of well over $1 billion every year. In 2015, Forbes announced that the LEGO group had overtaken Ferrari as the world's most powerful brand.

What's interesting about this breakthrough for LEGO is the lesson that agility on its own was not enough. LEGO was being highly agile in developing lots of new lines of toys. However, it was not working. The breakthrough came only when they used their intuitive intelligence (Pillar 4 of switch craft) and really tried to understand what their core audience wanted. This is the supercharged agility—informed by intuition and situational awareness—that is the essence of switch craft. Rather than merely trying to do something different, LEGO embraced new technologies and found new ways of engaging users of all ages to play and experiment with its small plastic bricks. Knudstorp's *informed* mental agility led to a completely new approach that transformed the company. The ability to stay agile, and create change rather than just respond to it, has been right at the heart of the LEGO revival.

Why can it be so hard to think differently?

Looking at these stories from the outside, it can seem easy to make this switch to agile thinking. So why do we find it difficult to change our mindset? The reason is that we have inadvertently practiced one way of doing things, or one way of thinking, over and over again, often from an early age, so it is very difficult to break away from this well-rehearsed method. The test below is a good example of how hard it can be to break away from habitual and well-practiced ways of thinking. You can test your ability to switch with this classic nine-dot problem. The task is to join up the nine dots below with four straight lines without taking your pen off the page:

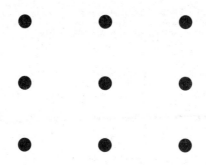

Despite its apparent simplicity, this puzzle is fiendishly difficult. Once you have seen the solution (Appendix 1 at the end of the book), it will become clear why. The problem is that the brain gets fixed in a familiar way of thinking, so that we look at the dots as if they were located at the edges of an imaginary square. We assume that we must stay within the boundaries of this square—there is even some speculation that this puzzle might have inspired the management phrase "thinking outside of the box" that became popular in the 1980s. However, when we realize that we are not restricted by the boundaries of the imaginary square, solving the puzzle becomes much easier.

As John Maynard Keynes, one of the most influential economists of the twentieth century, warned: "The difficulty lies not so much in developing new ideas as in escaping from old ones." Habitual ways of thinking can be very hard to shake and to a large extent this is because switching from one task or train of thought to another is not easy.

Opening our eyes and our minds

The discovery of bacteria is a great example of how this plays out in the real world. In medieval times, contagious diseases and epidemics were a constant threat. Fast-spreading maladies often appeared during the hot summer months, especially in densely populated areas, when the air filled with vile odors of garbage, animal, and human waste. It was thought that invisible vapors, which were released by rotting organic matter, could invade the body and play havoc with its vital functions. This "bad air," or "miasma," was the primary explanation suggested for the Black Death that killed up to two hundred million people across Europe in the mid-fourteenth century. Much evidence supported the miasma theory, which held sway until well into the nineteenth century. In 1864, however, a series of definitive experiments conducted by the French chemist Louis Pasteur overturned the miasma theory for good and replaced it with "germ theory." What we now know, of course, is that by removing the source of the bad smells, health reformers were inadvertently removing the real cause of disease, *bacteria*.

Remarkably, though, an Italian poet, physician, and scientist, Girolamo Fracastoro, had actually anticipated "germ theory" over three hundred years earlier. In 1546 he wrote a book, *On Contagion, Contagious Diseases and Their Cure*, in which he proposed that infections are not caused by "bad air" but by "seed-like beings" or "germs" that can be spread from person to person. While he believed that these germs were chemical substances that could evaporate and be diffused through the air, and not microorganisms as we now know they are, this was nevertheless a fundamentally new way of thinking about contagion. But it fell on deaf ears. The dominant theory of the time said that "bad air" was the problem. And so, the idea that a germ-like substance might be the culprit simply did not break through the scientific hierarchy.

Just over a century after Fracastoro's ideas were first aired, in 1677, a Dutch scientist, Antonie van Leeuwenhoek, observed germs directly when he invented a superior microscope. When looking at droplets of water under his microscope van Leeuwenhoek was amazed to see tiny organisms—which he called "animalcules." Still no connection was made with contagious disease, and so the implications of van Leeuwenhoek's observations were not fully appreciated until the famous experiments of Pasteur in France almost two hundred years later.

The way in which the scientific community took so long to shift from "bad air" to "germ theory" is a perfect example of how working within a collective and inflexible mindset can blind us to facts that do not easily fit within the dominant way of thinking. It took scientists over two hundred years to literally "see" what was obvious in van Leeuwenhoek's experiments. Just think how much faster progress would be if we could open our minds to all the possibilities.

If we look back through history, we can see that many of the great leaps in human knowledge have been based on a shift in the way we think about familiar things in new and unexpected ways. But an inability to switch from one way of thinking to another means that we either miss or ignore the implications of potentially valuable information.

We only see what we want to see

Opening our minds beyond the rigid constraints of our own values and beliefs can also have a powerful influence on what we observe about the world. We all tend to "see what we want to see."

This is perhaps never more apparent than when you are watching your favorite sports team. A famous psychology experiment conducted after an ill-tempered college football game in America demonstrated this beautifully. The year was 1951; the game was the Princeton Tigers versus the Dartmouth Indians. It was the last game of the season for both schools. Princeton had a star quarterback, Dick Kazmaier, who had featured on the cover of *Time* magazine that year; this was to be his last game. In the second quarter, uproar broke out on and off the pitch when Kazmaier's illustrious career ended with him having to leave with a broken nose and concussion after an especially rough tackle from a Dartmouth player. In the next quarter, a Princeton player snapped a Dartmouth player's leg. The match continued in this vicious back and forth, with Princeton finally winning 13–0. Tempers flared and accusations continued to fly long after the game had ended.

In the weeks that followed, the respective college magazines gave dramatically different accounts of the game. But psychologists from both Dartmouth and Princeton started to wonder whether it was possible that students from each school were genuinely "seeing" different versions of the game. To find out, they teamed up and recruited 163 students from Dartmouth and 161 from Princeton and asked them to watch a recording of the game and then fill out various questionnaires.

The findings were striking. Almost all the Princeton students (86 percent) and most neutral observers said that it was Dartmouth who started the rough play, while only 36 percent of the Dartmouth students agreed that their college had initiated it. When watching a film of the game, Dartmouth students were only able to pick out half of the rule violations made by the Dartmouth team. The contention is that people did not just *claim* to see different things, but that they actually did see very different versions of the game depending upon their college allegiance.

The study is often taken as evidence for a much wider truth. None of us are impartial observers of events. What we "see" is heavily colored by our own preferences and biases. This causes us to be quite inflexible in what we notice; only things that fit with our beliefs are accurately perceived. It's why we are much more likely to notice wrongdoing in strangers than in our friends. This can lead to very poor decision-making and wrong assumptions. Beliefs and allegiances can lead us to be close-minded and rigid in how we interpret what is going on around us.

As we've seen throughout this chapter, mental agility can transform our lives. It helps us to open our mind to different possibilities, see things more clearly, helps us to thrive and to succeed in sport, business, and daily life. Mental agility is made up of several different components, including some fundamental mental processes that help us to switch from one way of thinking to another, which we will explore in the following two chapters.

Chapter Summary

- A more agile mind will help you to thrive, make better decisions, and open your mind to a wider range of possibilities.
- Being agile is largely about being able to escape from old ways of doing and thinking about things.
- An agile mindset brings an advantage for business decisions as well as in your personal life.
- Mental agility is not about changing for the sake of it, it is an informed flexibility that chooses the right approach for the situation.

CHAPTER 6

THE NUTS AND BOLTS OF AGILITY IN THE BRAIN:

COGNITIVE FLEXIBILITY

I've only once been completely overwhelmed by fear.

I was twelve years old and living by the sea on the outskirts of Dublin. In the summer holidays my friends and I would spend most days at a nearby cove, scrambling down a steep cliff path that emerged onto the small sandy beach. A circular wall with a diving pillar that rose a meter or so above the wall had been built into the sea to create a large swimming area at low tide. At high tide this was all hidden, but as the tide went out, the pool and pillar would emerge.

On one warm but breezy day, we were splashing around in the pool when large waves began surging over the wall. The pool filled rapidly as the tide came in, leaving us quickly out of our depth. This was not unusual; I was a confident swimmer and enjoyed the crashing waves, until suddenly a particularly strong breaker caught me off guard and smashed me roughly against the wall. Winded and gasping for breath, I managed to scramble to the diving pillar and clung on as the waves crashed around me.

Looking around, I realized that the others had all managed to climb out of the pool and I was the only one left in the choppy water. The shore was about 30 yards away, and I could have easily swum to safety, but I became completely rigid, frozen with fear, and gripped on to the pillar for dear life. After what seemed like an age, a boy called James, who as it happened was one of the best-looking local boys, swam out to help me. I insisted that I couldn't move and would stay by the pillar until the tide went out. This was of course completely irrational—I knew it would take several hours for the tide to turn, and that the sea would rise well above the pillar before then.

James eventually convinced me to release my grip on the pillar and swim back to shore. I was shaken, but my friends had little sympathy. As I was a strong swimmer, they thought it must have been an elaborate act to entice the good-looking James to come and rescue me. "You deserve an Oscar," one of them laughed. To this day none of them believe me, but I had been genuinely frozen by fear.

Why is it so easy to "freeze"?

Many years later, as a neuroscientist and psychologist, I understand why I was so immobilized—the answer comes from our evolutionary history. Since predators can detect the slightest movement, one natural response to fear is to stay completely still. Vestiges of this reaction remain in our own brains, so we often freeze—even just momentarily—when afraid. But emotional reactions can be useless in the wrong context. Although "freezing" may save the rabbit that's about to be spotted by a fox, it is disastrous for the same rabbit facing the headlights of an oncoming car. Similarly, my brain's insistence that I cling to the pillar was clearly not a good option for the circumstances. But it's our basic biology that is to blame.

It's not just our physical reactions that can become inflexible, our thoughts and emotions can also become rigid. My own research has shown that a common thread running through many of our problems is a difficulty in disentangling ourselves from unhelpful feelings, thoughts, and actions. Think of the mental loop you can get into when faced with a persistent worry. No matter how much you try to distract yourself, your mind keeps coming back to that niggling thought. While your familiar ways of doing things may be comforting, it's important to ask yourself continually whether your approach is really a good fit for the problems you have.

"Cognitive flexibility" is the brain's bass riff of our agility

To maintain our psychological well-being and zest for life it's important to shift from rigidity and toward agility. To understand this, we need to first look deep into our brain where we will find a superfast process that goes on, largely outside our awareness, called "cognitive flexibility." This is the nuts and bolts in our brain of a capacity to stick and keep doing the same thing (easy), or switch and do something different (difficult). My decision to cling to the pillar reflected the "easy" option of sticking, while plunging out into the waves and overcoming my fear was a much more difficult option of "switching" my approach. While (initially at least) I didn't spend too much time thinking about it, this is an example of the benefits of the agile mindset that occurs in our conscious thoughts and actions. In this chapter, we go under the hood to take a closer look at the low-level, largely unconscious, processes that take place in our brain that support an agile mindset. These brain processes are called "cognitive flexibility" in the psychology jargon and they help us when deciding whether to "stick" or "switch."

What happens in the brain when you stick or switch?

Two different internal processes in your brain support these alternative options: "cognitive stability," which is the capacity to stick and persist in the face of distraction, and "cognitive flexibility," the ability to switch.

Cognitive stability involves two mental stages:

- First, maintain your focus on your current goal.
- Then, suppress all thoughts of the alternatives.

Cognitive flexibility is more complex and involves four different mental stages:

- First, shift your focus to a new goal.
- Next, suppress the old goal.
- Once done, then update your understanding of what you need to do to achieve your new goal.
- Finally, put into action whatever is required to achieve your new goal.

These two capacities have more in common than you might think. Brain imaging shows that they are both influenced by the same brain regions within the frontal cortex—an area of the brain that is vital for important cognitive functions. I was initially surprised by this overlap, but after some thought it seems obvious that whether you are trying to block out all thoughts of alternative possibilities (and stick) or inhibiting an old goal (in order to switch) your brain needs to suppress lots of different thoughts and actions. So, it makes sense that this ability to suppress (known as "cognitive inhibition") is just as vital for agility as it is for grit.

What brain imaging studies also show is that agile people have more flexible connections among different regions of the brain, similar to what we observed with our teenage volunteers. A more agile brain is fluid and connections can be dynamically reorganized in the moment to support any mental process that is urgently needed. It's important to appreciate that these connections are not hardwired in any way—all of us can loosen up our brain networks by training them. If we work on helping our brain to switch rapidly from one thought to another, then this will help us to become more agile in our everyday behavior.

Cognitive flexibility

In our brain, cognitive flexibility refers to the low-level process that occurs in the brain to enable us to switch from one task, or what we can call one "mental setting," to another. If you take a sip of water, for instance, picking up the glass is one setting your mind is fixed on, moving it to your lips is another, and then sipping and swallowing is yet another. In terms of brain processing, each of these different components in the sequence requires a switch from one mental setting to the other. My argument is that this ability to switch fluently from one mental setting to another not only helps the fluidity of our behavior, it also underpins our ability to overcome habitual ways of thinking and to transition from an old way of thinking to a new way.

As we saw in the previous chapter, all of us sit along a spectrum that ranges from extreme rigidity to easy agility, and as we grow older we often become less flexible and more set in our ways. However, before we can develop our agility in this broader sense it's important to work on the nuts and bolts: the fluidity of our brain-based cognitive flexibility. We can *learn* to become more cognitively flexible, and developing the fluidity of this transition—from one mental setting to another—is essential to help us adapt.

How we learn to "task switch"

Between the ages of about seven and eleven, the ability to switch the mental settings that allow us to switch from one task to another is something that develops naturally. A good way to test this is to ask children to sort a deck of picture cards in different ways. Let's imagine a pile of cards with pictures of both animals and sweets in either blue or yellow. Most seven-year-olds can easily sort the cards into two piles based on "yellow" and "blue" or "animals" and "sweets," but they will struggle if they are then asked to switch their mental setting and make one pile of, say, blue animals and another pile of yellow sweets. However, by the time they are eleven most children will find this easy.

This ability to "task switch," as psychologists call it, is important for everyday life. For instance, switching between the idea that an engineer can be "male" or "female" is tapping into a very similar cognitive process. Indeed, several studies tell us that children who are good at this task are less likely to use rigid stereotypes about people and are also better at developing skills such as reading. So, this is an important low-level cognitive process that we can work on improving to help us develop our agility and psychological functioning in a much more general way.

Test your own ability to task switch

Cognitive psychologists like myself use a range of task-switching tests that are essentially a more sophisticated version of the child's sorting task just described. In a nutshell, task switching allows us to quantify the momentary disruption caused by switching from one mental setting to another.

For example, consider a series of digits in either bold font or in normal font to which the following rules apply:

- If the font is in bold (e.g. **7**) you have to say whether the digit is higher or lower than **5**.
- If the typeface is normal (e.g. 4) you have to say whether the digit is "odd or even."

In technical terms, one mental setting is "higher or lower than five," while another is "odd or even." Switching between the two disrupts the smooth flow of mental processing. For instance, the answer to the following sequence—6 2 7 **4 8** 3—involves two "repeats"—"even" "even" "odd," where the mental setting is the same "odd or even"—then followed by a switch from "odd or even" to "lower or higher," followed by two final repeats—"lower" "higher" "lower." The switch in the middle involves a switch from one *mental setting*—"odd or even"—to another mental setting—"higher or lower than 5."

Try this out for yourself with the digits presented below.

- Open the stopwatch on your phone and time yourself from the beginning to the end.
- Remember, normal typeface is "odd/even," **bold** typeface is "higher/lower than 5."

6	2	7	4	9	3
6	3	8	3	2	9
1	**3**	**4**	**8**	**6**	**6**
7	**4**	**8**	**2**	**3**	**9**

Fill in your time here: _____

The first time I did this it took me 21.32 seconds.

Now, reset your stopwatch and do exactly the same with the following sequence—again, time yourself from start to finish. The rules are the same—normal typeface = "odd/even," **bold** typeface = "higher/lower than 5":

6	2	7	**4**	9	3
6	**3**	8	3	2	**9**
1	3	**4**	**8**	6	6
7	**4**	**8**	**2**	3	**9**

Fill in your time here: _____

My time for this one was 26.88 seconds, meaning there was a *switch cost* of 5.56 seconds. The second set is harder because several of the sequences are *switch* trials, while there is only one switch in the first set of digits. How did you do? If you practice your task switching regularly you should see an improvement over time.

You can do this by coming up with your own numbers and randomly printing half in boldface. If you do this regularly it will become easier, and it will give your brain a good cognitive flexibility workout. However, it's also important to target your cognitive flexibility with more everyday tasks, as we will see below, rather than just sticking to one technique.

Training your cognitive flexibility in day-to-day life

Cognitive flexibility is a vital brain process that supports agility and flexibility in everyday life. For instance, many everyday situations such as taking up work again after a break, winding down on a vacation after an intense period of work, or simply shifting from one activity to another, require cognitive flexibility to operate smoothly. Here is a simple exercise to improve this fundamental aspect of your mental agility:

1. Write out three or four tasks that will take no more than about 10 to 15 minutes each to do. They can be something like writing a short email, making a phone call, booking theater tickets, or tidying up your desk.
2. Work out a sensible amount of time for each of your activities and decide on the order in which you will tackle them.
3. Now, set a timer for your allocated time and begin the first task. When that time is up, stop. No cheating, no matter where you are in your task, even if you have almost finished; you must stop as soon as the buzzer sounds.
4. Take a short break. Then reset your timer for the appropriate time and get on with your second task.

This simple assignment is surprisingly helpful. Firstly, it will help you to find out how good you are at estimating how long certain tasks will take. Hint—most of us hugely underestimate how much time simple tasks, like sending an email, actually take. Second, you will also learn to switch more efficiently from one task to another. If you do this regularly, perhaps once a week, you will dramatically improve your cognitive flexibility—the bedrock of your broader agility.

More advanced versions of this exercise involve setting a timer to go off at random intervals. In some studies, people are given three tasks to do over a thirty-minute period. A timer is set to buzz at six random times during that period. If the buzzer goes, the person must instantly switch to the next task. There's no break this time because you are training your cognitive flexibility directly rather than trying to be more efficient. So, you must stop what you are doing instantly and shift to the next task. Done regularly, this exercise will do wonders for your brain's powers of agility. It's also a great way of clearing a few chores that you have been avoiding.

Multitasking drains your energy

While shifting from one thing to another is a great exercise to boost your cognitive flexibility, it is also an important reminder that switching like this takes up energy and effort. Remember that multi-tasking is largely a myth—what actually happens is that we rapidly switch from one task to another. So, if at all possible, try to plan your time so that you can concentrate on one thing at a time. Flipping back and forth between different tasks is very draining. I'm often guilty of this, stopping to check out an email that's just come in when I am in the middle of something else. It's hugely distracting and an inefficient use of time. As I am writing this, for instance, I have switched off all email alerts and other notifications. So, if you have several tasks to complete during a morning or a full day, be disciplined about planning your time and try to focus on just one thing at a time. Not only does good time management support your well-being, it will also give you the energy and focus required to perform at your best.

A good starting point is to decide on a couple of things that you need to do in a day. Once you have made this decision, be rigorous about allocating a certain amount of time for each task—and be realistic. This requires discipline. Try to follow this routine on a regular basis:

1. Start the day with a plan to undertake two or three tasks. It might not sound like much, but any more than three tasks in a day and the cost of shifting your mental settings will start to really eat into your efficiency and energy. The tasks should be specific rather than open-ended. So, rather than having a vague intention to "work on my book," I might "complete a specific section in Chapter 2." It's important you define the parameters of the task so you don't end up feeling like you've failed when what you set out to do wasn't possible.

2. Once you have chosen your two or three tasks, you then need to prioritize them in order of importance or urgency. If one of them must be finished that day, then perhaps that should be your only task for the day depending upon how long it is likely to take. Again, be realistic and don't put yourself under unnecessary pressure.

3. Now, give yourself a sensible time frame to complete each of your tasks. To begin with you will probably vastly underestimate the time a particular assignment will take. With practice, however, you will become better at figuring out how long something will take. Because we know that switching from one activity to another requires energy, make sure to account for this by having a break of at least fifteen minutes between tasks. This vital gap will help to disengage your mind from the first assignment. Only then can you truly shift your mental setting

to the next task and get on with that. If you adopt that principle consistently you will not only end up being more efficient but also have much more energy at the end of each day.

It's also important to schedule in rest periods, opportunities for exercise, and time to check emails. Like most academics, I regularly receive about 150 to 200 emails each day and find my inbox can be overwhelming. The only way to manage this is to spend an hour in the morning and the evening working through the most urgent of them. I'm not always rigorous about this—and when I'm not I suffer for it, as hours can easily be soaked up by emails and then I end the day feeling stressed and frustrated because I have not completed the things that I wanted to do.

If you want to go to the gym, get out for a run, or have a yoga session, schedule a time—and stick to it. You may have to get out of bed an hour earlier in the morning, but it's important to schedule in a time and then be disciplined in carrying out your plan.

Video games and travel may also help

Playing fast-action video games also involves switching rapidly between multiple rules, actions, goals, and targets. While there is a mixed bag of results, there is some evidence that playing these games can lead to more efficient flexibility.

Another way to boost your cognitive flexibility is to travel. The mind's ability to switch efficiently between different ideas is a cognitive ability that also helps creativity. One research team examined the creativity of senior designers at 270 high-end fashion houses. Those designers who had lived in several countries consistently produced more creative fashion lines than those who had not. Further investigations revealed that the type of country matters; living somewhere that had a very large cultural difference from your own did not lead to creativity boosts. One explanation for this may be that it's harder to engage with a very different culture, especially if you don't speak the language.

This suggests it is immersion and engagement with a new culture that really makes a difference to your mental agility. Cross-cultural experiences can pull you out of your cultural bubble and give you an enhanced sense of connection with people from different backgrounds. As Mark Twain once commented, travel "is fatal to prejudice, bigotry, and narrow-mindedness." And it's also a great trainer for mental agility.

Find unusual uses for common objects

The "Unusual Uses Test" gives you a limited amount of time to think of as many uses as possible for everyday objects, like a tin can, a cup, or a paperclip. This gives an indication of a person's fluency, creativity and mental flexibility. You can practice this at any time by looking around the room or train or plane or wherever you are, choosing an object and then seeing how many uses you can come up with for that object. If you have children, try it with them too—it can be fun and, done on a regular basis, will help to improve your flexibility and enhance creativity and agility for all of you.

Practice is important—especially if you are anxious

A fundamental feature of our brain is to find task switching disruptive. It will always be easier to keep doing the same thing rather than switching. This is why productivity gurus often tell us to avoid unnecessary switching between tasks and being interrupted by things like emails during the day, because it can dismantle our concentration and undermine performance. While we will always be disrupted by task switching, anxiety will cause us to struggle with it to an even greater degree. Anxiety also makes us more sensitive to the types of tasks that we are switching between.

Imagine you are absorbed in writing a difficult report, or solving a complex problem, something that requires intense concentration. You then have to switch your attention momentarily to a simple chore such as making a restaurant booking before returning to your task. When we are not particularly anxious, we find it equally difficult to switch away from easy and difficult tasks. However, as our stress levels increase, we find it more and more difficult to drag our attention away from a very engaging task compared with switching from an easy task.

Pulling our minds away from all of the internal chatter that anxiety involves means that we have to make more effort to reach the same level of performance compared to when we are less anxious. This is why externally you can't see any difference in performance between groups of highly anxious and more laid-back people. But internally a different story emerges: the brains of the anxious people are working harder to achieve the same level of performance. Like swans paddling under water, it looks elegant and easy from above but under the surface lies a different story.

Anxiety undermines our cognitive flexibility— and our enjoyment of life

My own work on anxiety has shown that by biasing our cognitive processes, anxiety can alter what we are conscious of and can distort the way in which we experience reality. However, this tendency to distort reality and see potential danger lurking in every corner is not the real problem. In genuinely dangerous situations this is a perfectly appropriate cognitive mechanism that operates deep in our brain and protects us. The problem is that when we are regularly anxious this tendency becomes our default, and we lose the agility to try out different ways of seeing things. Anxiety effectively puts up a roadblock in our mind.

Anxiety undermines the fluidity with which we interpret the world around us and instead immobilizes what should be a very dynamic and agile system. An anxious brain causes the entire body to get fixated on what might go wrong rather than focusing on flexible ways to solve a problem, and often results in highly rigid repetition. We are quite literally stopped in our tracks and the constant awareness of threat disrupts how we think, how we feel, and how we act.

While some studies show us that anxious people have difficulty with task switching, what has been missing is a study following people over a period of time. As every psychologist will tell you, a correlation, or association, does not imply causation. Anxiety might lead to cognitive inflexibility just as easily as the other way around. What is called a "longitudinal design," in which we follow a group of people over time, is helpful because it allows us to figure out whether a degree of mental arthritis really does lead to problems in dealing with the ups and downs of everyday life. Is it the case, I wondered, that a difficulty in switching between tasks would predict increasing stress and worry over time?

Instead of using the traditional version of task switching that uses neutral items, such as digits, for this research project we decided to use an emotional version. Given my own earlier findings that anxious people tend to get stuck on negative material, it seemed likely that anxious people would find it especially difficult to switch away from emotionally charged, especially negative, information. Using this new switching task, that was exactly what has been found. An inflexibility to switch away from negative material is indeed associated with a tendency to respond to stress with a potentially dangerous, and ineffective, coping mechanism—negative rumination.

This gave us the impetus we needed to conduct our own study. We decided to test a group of students over an eight-week period. What we wanted to find out was whether those who were more cognitively inflexible, especially with emotional material, would respond to daily hassles with increased worry (which is a repetitive, toxic, and often ineffective way to deal with stress).

To better understand how people coped with everyday pressures we used the appropriately named "hassles and uplifts" questionnaire. Each volunteer filled out this questionnaire online every week, providing us with a cumulative record of the number of hassles (e.g. missing the bus, being late for work) and uplifts (things like meeting friends, positive appraisal at work, or a good essay mark for students) that they experienced each week. At the baseline testing session, our volunteers were presented with a positive or a negative scene (a couple gazing lovingly into each other's eyes or a couple arguing, for instance) and asked to classify the image as quickly as possible (via a button press on a computer) according to two different sets of rules:

- An "emotionality" rule—is the vibe of the scene positive or negative?
- A "numbers" rule—are there two people, or more or less than two people, in the scene?

The idea was that when you have to shift, with just a split-second's notice, from a decision based on "emotion" to one based on "numbers," this would reveal a switch-cost and give us an insight into cognitive inflexibility for emotional images. Our expectation was that our volunteers who showed a high degree of *inflexibility* in switching away from the negative emotional aspects of a scene would report higher levels of anxiety and worry over time. The brains of these subjects would be less able to switch away from threat compared to those who were less prone to anxiety.

The entire experiment took about six months to complete and the final results were intriguing and not at all as we expected. While there were some signs of increased rigidity in shifting away from negative scenes as we had predicted, it was the degree of rigidity in switching *toward* the positive emotional aspects of a scene that turned out to be most important. Anxious people were less efficient in switching *toward* positive images. What was remarkable was that these less-agile individuals were also far more bothered by the hassles they had experienced, and did not seem to gain as much benefit from the number of uplifts they had experienced over the entire two-month period.

A simple laboratory measure of emotional task switching really did seem to predict anxiety and worry in everyday life. This is an important and novel finding because we measured emotional task switching first, and then examined what happened across the following two months. The results were telling us loud and clear that cognitive inflexibility, at least in terms of how people process positive situations, can undermine mental well-being and is associated with significantly higher levels of worry.

This rigidity of mind, or mental arthritis, leads eventually to highly stereotyped ways of responding so that people become more and more divorced from reality. What happens is that their mind tells them that everything is getting worse, or that nothing works out for the best, or that if something can go wrong it will, or all three at once. Cognitive biases like these ensure that people often overlook the good around them and fail to remember the positive things that have happened. When anxiety escalates, these psychological mechanisms, although useful in adversity, cannot be switched off when things are going well. Instead, they become inflexible and rigid, and undermine our psychological vitality.

Whether we feel anxious or not, there is always some cost to switching from one task to another. However, the more anxious we feel the more deeply absorbed we tend to become with a task. And this can lead to more difficulty in shifting away the more difficult the task becomes. So, while we should try to avoid switching between tasks as much as possible to help our productivity, when we are feeling anxious or stressed it becomes even more difficult and disruptive—unless, that is, we are practicing task switching.

Task switching is not just useful to keep cognitive psychologists in business, or for children's games, or for solving fun puzzles. It is a fundamental brain process that underpins much more sophisticated and complex decisions that we make in our everyday lives.

Cognitive flexibility is the source of our capacity to be agile

We have seen how switching from one thought, or mental setting, to another comes with a cost. Whether we are young, middle-aged, or old, a difficulty in switching acts as a mental block that stops us solving problems, making good decisions, and even seeing the world in an objective way. While cognitive flexibility is something that plays out in our brain, it does not stop there. It supports a much broader agility that influences how we feel, how we think, and how we act in the world. Psychological agility in this wider sense is now a thriving area of research and many studies have found that this capacity is a cornerstone of our psychological well-being.

It's important to remember that our capacity for agility is rooted in very deep-seated processes in the brain that allow us to switch from one thought or activity to another. The only way to train this cognitive flexibility is to practice and practice and practice. Just as a an athlete will hone their craft by endless training, so each of us can practice task switching on a regular basis. The benefits are profound. Not only will it support our mental well-being, as my own studies have shown, it will also bring us a range of other benefits to navigate an unpredictable and complex world.

Chapter Summary
- The ability to switch is more complex than the ability to stick and therefore requires more effort.
- Cognitive flexibility is the fundamental brain process that underpins a broader sense of agility.
- To improve this broader sense of agility, which is the first pillar of switch craft, we need to first work on our cognitive flexibility.

- Psychologists measure cognitive flexibility by means of task-switching tests and these tests can also be used as training exercises to improve your cognitive flexibility.
- Your level of anxiety can influence the flexibility of your brain processes: less anxiety = more flexibility, more anxiety = less flexibility. This is especially the case when dealing with emotionally charged material.
- Being able to avoid distraction is just as important for deciding to switch as it is for keeping focused and concentrated on a task.
- Cognitive flexibility is something we can practice and improve.

CHAPTER 7

THE ABCD OF MENTAL AGILITY

A couple of years ago I joined a team of psychologists delivering a workshop designed to help serving police officers deal with stress. One officer, Mark, told us a story about a situation that he'd had to face early in his career, soon after he had completed his training. Early one evening a call came into the police station reporting a domestic argument that was taking place on a suburban street. A neighbor was concerned that things were on the verge of taking a violent turn. As it turned out, the police had received many calls from this same street to deal with a married couple involved in loud and violent arguments that usually ended with the man kicking doors and then storming off.

While disturbing, these arguments had never resulted in physical violence. Assuming that he would be faced with the same couple, and the usual scenario, Mark set off expecting that he would have to try to calmly defuse a drunken feud between husband and wife. His training had prepared him well for just such scenarios. When he got to the scene, however, Mark was completely unprepared for the chaotic situation with which he was faced. Instead of the couple he had been expecting, several members of the public were standing around in shock. In their midst, a heavily bleeding woman was lying motionless on the ground. Nearby a man was bleeding heavily with bad bruising on his face, and a third man with serious injuries, still shouting, was being held back by two members of the public.

Mark's initial reaction was to freeze. As people looked toward him, expecting him to take control, he quickly came to his senses. Having established that an ambulance had been called, Mark then began to try to get to the bottom of what had happened. The man and woman, he gathered, had been arguing, a bystander had become involved, and this intervention had led to a violent fight between all three.

Mark's initial reaction was typical of what happens in dislocated expectation. His brain went into a kind of microfreeze as he tried to first undo the strong expectation that he'd had. Mark then had to process what actions he needed to take so that he could deal effectively with the new situation. This experience illustrates the essence of mental agility. First, we need to disengage from our initial expectation—our old mental setting—and then shift to a new setting and think about the most appropriate course of action for the new circumstances.

An appreciation that the unexpected happens is essential to allow us to perform at our peak in any situation. We can think of it in the same way as planning an expedition to a remote part of the world. You might begin by finding out as much as you can about the kind of terrain you are likely to encounter. To get over high mountains, perhaps in snowy conditions, to navigate through rivers, to traverse hot dry deserts, all require different skills and different equipment. Life is not dissimilar. As we progress through our life journey, we must learn how to cope with parents, siblings, friend-ships, puberty, conflicts, marriage, starting new jobs, leaving jobs, illness, death of close friends, the changes that come with our own aging, and many other challenges. Sometimes changes can be predicted while at other times they are foisted upon us unexpectedly. Either way, our openness to accept and cope with what are technically called "dislocated expectations" is the cornerstone of our psychological health.

Agility, which is the first pillar of switch craft, is not just about the big decisions in life—it also operates in the smaller decisions and in our everyday activities. As a child growing up in Ireland, my summer holidays were filled with traveling around the country on the junior tennis circuit. Almost every week there was a tournament in some part of the country. We were a tight band of players who knew each other's styles of play very well. It was fun and fiercely competitive. Gemma, for instance, was one competitor whom I frequently played. Our matches were invariably close, but she beat me more often than not.

My best win against Gemma came in the quarterfinals of the North Dublin Championships, at my home club of Sutton Lawn Tennis Club. I can still remember the rare excitement of achieving a semifinal place in a major tournament. But it almost didn't happen. Gemma was unusual in that her backhand was much stronger than her forehand. In the first set, I played my usual strategy of plugging Gemma's forehand side. Play was not going well, Gemma hardly made an error, and forehand after forehand fizzed past me. Occasionally, the ball drifted to her backhand where, surprisingly, she made several uncharacteristic errors.

Did I notice? No.

But when I was sitting at the change of ends, having just lost the first set, my coach whispered that I should change tack. "Aim for her backhand," he said, "it's looking shaky." For the next few games I just could not do it. Keeping away from Gemma's usually powerful backhand was a tactic so deeply ingrained in my mind that I found it almost impossible to adjust. Thankfully, I finally forced myself to focus on Gemma's backhand and the tide began to turn. I won the second set, just. Then I went on to win the final set and achieved that rare semifinal place.

For much of that match, I had been blind to what my coach, Aidan, could see so clearly. His advice at the end of the first set had pushed me to do the obvious: to respond to the situation at hand, rather than relying on games past. My inability to shift away from a tried and tested strategy—my lack of mental agility—had almost let me down.

While cognitive flexibility is the bedrock of our mental agility in the brain, as we saw in the previous chapter, our mental agility plays out in a much broader way in terms of how we act, think, and feel in everyday life. To become agile, psychological research has shown us that there are four key elements that we need to work on.

The four elements of agility

Broader psychological agility consists of four dynamic processes, each of which unfolds over time. I have called these the "ABCD of agility" and they reflect how well a person can:

- **Adapt** to changing demands
- **Balance** competing desires and goals
- **Change** or challenge their perspective
- **Develop** their mental competence

To be agile, we must nurture these four elements of agility as illustrated in the following diagram:

The ABCD of Agility

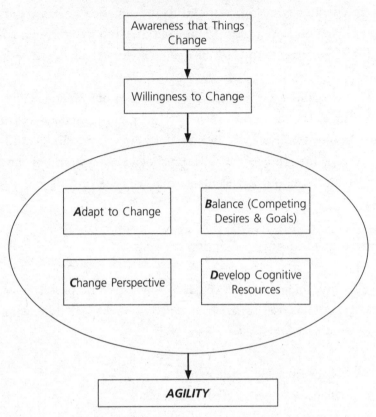

Adapt to changing demands

One group of people who experience high levels of change and uncertainty are military families. They typically move every two years or so, sometimes to another country. In April 2017, I gave a keynote talk at a congress on resilience and adaptability in Arlington, Virginia. While there, I met a group of women from the Army Wife Network who told me how difficult it was to raise happy, well-adjusted kids under these circumstances. Military kids must become used to leaving their friends, and learn how to adjust to new schools, new friends and new locations. I asked the women how they coped. Their main advice was to emphasize the positives of constant change and to focus on opportunities rather than problems. While acknowledging the difficulties, focusing on the chance to travel, to meet new people, and to learn new languages can make these changes feel exciting.

This advice resonates with the wealth of research that shows how training actively to adapt to change can be transformative. Being adaptable is essential simply because circumstances, for other people as well as ourselves, are always changing. We have seen in the early part of this book that while we often resist change, our biology and our psychology are both designed to help us adapt.

Being adaptable doesn't mean acting on impulse without a plan. True adaptability begins with a strategy, continues with a deliberate action, and involves letting go and moving on. In business, if the market has changed, you need to adapt or you will be left behind. This means that well-worn habits may need to be broken if the situation demands it.

A Zen parable tells the story of an old monk who is traveling with a younger monk when they come to a river with a very strong current. At the riverbank, a young woman is waiting and asks for their help to get across. The two monks glance at each other, as they have both taken a vow to never touch a woman. Then, without any further hesitation, the older monk picks up the woman, crosses the river, and sets her down gently on the other side. The monks continue their journey until several hours later the younger monk can no longer contain himself. "Why did you carry that woman when we took a vow not to touch women?" he asks. The older monk replies, "I put her down hours ago at the side of the river. Why are you still carrying her?"

To me this is a wonderful example of adaptability. It's a reminder that we should not dwell on the past in a way that affects how we live in the present. As the British entrepreneur Richard Branson has said: "Every success story is a tale of constant adaptation, revision and change." He believes "a company that stands still will soon be forgotten."

How to improve your ability to adapt

Adaptability is something that can be practiced. Getting used to change and new situations regularly is an important part of this. I'm not expecting you to orchestrate a major relocation every couple of years like the military families, but do look for opportunities to try new things that will help you learn new skills. Remember that adapting is not something you do every now and again, it is a continuous and constant process. The following habits that help adaptability should become a lifestyle rather than being called upon only when there is a crisis.

- **Nurture a curious outlook:** Ask yourself lots of questions about an upcoming change. Is it positive? What are the opportunities? What are the downsides? What might be lost?
- **Make sure you have a range of options:** Make sure that you have a Plan B and a Plan C; it's always risky to just have a Plan A.
- **Take other people's concerns into account:** Don't dismiss people's worries without listening to them, especially if you are in a leadership role. Consider mentoring someone else who seems a bit stuck and think about how you can help them to be more open-minded and adaptable.
- **Take care of yourself:** Even if you are very adaptable, don't forget that change always takes some toll, so make sure you have a good support structure around you. Look to friends, trusted colleagues, mentors, and others to support you if you are going through a difficult change.
- **Regularly expose yourself to new situations:** This is vital to do on a regular basis, not just when you are confronted with a big change. By regularly engaging in new activities and meeting new people, adaptability will become a habit.
- **Diversify your life portfolio:** Or, in other words, don't put all your eggs in one basket. Investing your time and effort

in a diverse range of activities, life roles, and experiences is an essential ingredient to increase your adaptability. Exposing yourself to multiple life experiences and social roles enhances your confidence.

- **Surround yourself with extraordinary people:** The science shows us that watching someone like ourselves do something extraordinary encourages us to think that we could give it a go. So, to improve your adaptability, try to nurture a personal portfolio not only of skills and experiences, but of people who can support you in different situations.

Balance competing desires and goals

Each of us has many competing goals and priorities and things we want to do. The problem is that we are limited by time and energy— there is only so much that we can achieve each day. Do we dedicate that day's energy and time to excelling at work? Or do we use it to be a good friend and look after someone we love? The reality is we can't be all things every day and we must be able to balance and prioritize our goals and desires.

Equally, despite having many roles, we are one and the same person and the different areas of our lives are often closely interconnected. This was perhaps never more apparent than during the coronavirus pandemic when many people were working from home rather than having separate spaces for work life and home life. Many studies tell us that what happens at work flows into our home life, just as conflicts or excitement at home can spill over into our working life.

A good example is a study by a team of Spanish and British researchers who asked 160 people to keep a diary of conflicts at home and at work over a full working week. Unsurprisingly, what they found was an intimate relationship between conflicts at work and at home. If you had an argument with your spouse in the morning, the number of conflicts with colleagues at work increased during the day. This was a two-way street: if people had difficult interactions with colleagues at work this tended to spill over to further conflicts with their family when they got home. This study and many others like it remind us that we must be considerate about the emotional baggage that we bring home from work. While it's important to confide in your partner about the problems you may be having, be aware that the negative feelings caused are highly contagious. Good news is also contagious, of course, so it's great to bring some positivity home.

How to switch off from work

One way to do this is to remember the "fertile void" that we talked about earlier. Find a way to insert a gap between your work and home life. This will help you to leave the problems at work behind and arrive home in a more relaxed mood as a result (or, if you work from home, will enable you to switch "modes"). Before I moved to Oxford, I was head of a large and busy psychology department at the University of Essex. With over one hundred staff and eight hundred students, the demands of this leadership role far exceeded my time and energy. I often found myself sitting at home in the evening thinking and worrying about various issues that had come up at work, all of them seemingly urgent. I finally found a way to switch off from work and arrive home in a better frame of mind. Rather than driving the short distance home from the university, I would walk along a two-mile river path. Hail or shine, I always found that after about ten minutes of walking I would begin to notice the swans on the water, the birdsong in the trees, the smell of the river, and gradually the worries of the day would begin to fade away.

I was lucky in that the river path was available. But try to find some way of breaking the psychological link between work and home, ideally doing something that you really enjoy. Perhaps you could go to the gym or the pool, or meet a friend for coffee, to draw a line between the end of your working day and your home life. Other things to watch out for, which we all know but often don't do, is to keep away from your work emails in the evening. It's hard to disengage your mind from work if you are still reading emails and thinking about what needs to be done the following day.

A surprising way to switch off from work is to get involved in some voluntary work. Even though it might take up considerable time, volunteering has been shown to be highly restorative because of the new people you meet and a feeling of giving back something to your community. One study asked 105 German employees to keep a diary over a two-week period, and found that those who spent more time on volunteer work activities were better able to disengage from their normal work-related activities, and their mental well-being benefitted as a result.

Manage your time well to keep your life in balance

Finding ways to detach yourself from work is all very well. However, the reality is that many of us are swamped by the demands of work, so it's important to find ways to manage your time well. It also helps to be very clear about our main goals in life.

We all have various plans, goals, and desires, and it's important to integrate them in a coherent way. You may have a longer-term goal as well as several shorter-term goals that may or may not support each other. All of this creates pressures and imbalances that can lead to serious stress and exhaustion.

However, don't lose sight of the fact that we do have some degree of control over how we spend our time. Your time is your most precious resource, and while it's often difficult to escape social expectation, it's important to learn to use your time well and spend time engaging in valued activities.

One way to use time well is to regularly push the boundaries of your comfort zone. Try to find that sweet spot where you are challenged and stretched but not overwhelmed by a task. The very best moments in life come when your body and mind are stretched to their limit—but not beyond. This state of being completely absorbed in a task that is challenging but worthwhile is called a state of psychological "flow." This is when you are completely "in the zone," totally immersed in the pleasure of what you are doing, less aware of yourself and your problems, using your skills to the utmost, and ready to grow and develop.

There are three key questions that can be helpful in deciding how you should be using your time:

- *What* are your most important goals? These are the type of goals that are most often career-related. Write down some of these goals in your journal and make them as tangible as possible. What type of job do you want to do? How much do you want to earn? How much money do you want to save? Is there any conflict between your goals?
- *Why* do you want to pursue these goals? Many of us get stuck in a routine of overworking, forgetting what we wanted in the first place. So, step back and ask yourself, why do I want to achieve these goals? This question will often turn up more personal goals surrounding your family and your well-being. Perhaps you want to earn or accumulate a certain amount of money in order to support your family or to give yourself the opportunity for more leisure time. These are, of course, the most important goals of all in terms of boosting your happiness and well-being, so these are the ones that should be given priority.
- *How* are you going to achieve your goals? Once you have a clear idea of both your career-related and your personal goals, you then need to set out a clear agenda of how you

are going to arrive at the place you want to be. It's important when setting a goal to break it down into manageable and *measurable* parts. This will make your goals seem more achievable. It's also important to commit to your goal and perhaps tell someone about it, as we know we are more likely to stick with our goals if we have shared them with others.

There is a lovely Brazilian folktale that tells the story of a happy fisherman. Every morning he goes out in his small boat to catch enough fish for his family to eat. Then he comes home and plays with his children, has an afternoon nap with his wife, and then meets his friends in the village square in the evening. One day, a businessman comes along and advises him to get a bigger boat. "You could catch many more fish," he suggests, "perhaps eventually get a fleet of boats, employ more people, and become a rich man." "What would I do then?" asks the fisherman. "Well, once you have enough money you could have time to play with your kids, and spend time with your wife and your friends." The fisherman is puzzled. "Isn't that exactly what I'm doing now?"

It's easy to lose sight of the *why*, as this rich businessman had. Many of us are instinctively dissatisfied by the balance of how we spend our time and have often lost sight of our primary goals. That's why it can be helpful to take a step back and reflect on the reality of how you spend the time you have available; ask yourself if this is an effective allocation given your main goals, and what your ideal allocation would be. This was shown in one study, in which students were asked to allocate the ideal number of hours they would like to spend on particular activities on a daily basis, with the goal of achieving a better balance across life domains and interests. Each student was encouraged to move toward this goal and to write down specific goals that would help. Four weeks later, the students who had followed this schedule reported greater balance and happiness in their lives.

How do you spend your time?

There's a big difference between how you *actually* spend your time and how you would *like* to spend your time. The perfect synthesis will depend on your values and goals and will, of course, change over the course of your life. If you have just started your work life and are highly ambitious, it often makes sense to dedicate more time to work and relatively less to friends and family. However, when you have young children, you might now want to spend more time with family rather than in the office. Considering your current demands, it is important to try your best to prioritize those things that will get you closer to your most cherished goals. A good first step is to figure out how you actually do spend your time each day on average.

In the table below there are ten areas of life. You have twenty-four separate hours to allocate between these ten areas. To complete this task, first think about how you currently spend your time. Then, indicate how you would ideally like to spend your time.

Activity	How Do You Spend Your Time?	How Would You Like To Spend Your Time?
Being asleep		
Education of some sort		
Your main job/work		
Housework (e.g., cooking, cleaning, repairs, buying groceries, etc.)		
Volunteering (doing community work or service)		
Relaxing (playing sports, watching TV, playing games, etc.)		
Commuting (time spent traveling for work or school)		

Activity	How Do You Spend Your Time?	How Would You Like To Spend Your Time?
Relationships (spending time with friends, partner, or family)		
Looking after yourself (exercising, eating, bath, etc.)		
Spirituality (attending religious or spiritual service, meditating)		
Other (add another activity)		
Other (add another activity)		
TOTAL TIME UNITS	24	24

How to create more time

If you are like most people, you will find that there is probably a gap between your ideal allocation of time and how you spend your time. You probably spend more hours working than you would ideally like to, for example. In a survey conducted by the Mental Health Foundation in the UK in 2014, a whopping 58 percent of employees said they felt irritable when working long hours, 34 percent felt anxious, and 27 percent felt depressed. We all have pressures and commitments that are difficult to avoid, but with some careful self-management there are life hacks that will help you to buy more time to focus on your cherished goals.

Try out the following to get you started:

1. **Learn to say "no":** This is probably one of the hardest things to do. Like many academics, especially female academics, I often say "yes" to requests to review articles, to mentor students, to take on administrative roles in my department, to review grant

applications from research-funding organizations, to give talks, to provide expert advice, to sit on interview panels, to join university committees. The list is endless. If you never say "no," people will keep asking and you will soon get overloaded with things that are not actually part of your core work. This will inevitably lead to frustration, stress, and resentment. So, learn to say "no." There is no need to be rude about it and you don't have to justify your reasons other than to say that you do not have enough time. When someone asks you to do something, don't answer right away without thinking. Say you will get back to them, and give yourself some space to think about whether the request will take up too much time and whether it fits with your goals. Unless there's a very good reason to accept the request, politely say "no."

2. **Work smarter, not longer:** Prioritizing our time well will increase productivity. So, avoid multitasking—you now know how draining task switching can be. You may feel like you are getting lots done but it's a false impression. Follow the advice we saw earlier and focus on two or three tasks a day, no more, and be rigorous about protecting your time. One executive I coached realized that much of his time was being soaked up by one particular colleague who regularly wanted to talk. The solution was to meet this person for coffee for a fifteen-minute period twice a week, give them his full attention for that time, and politely and firmly say that he had to focus on other things until the next time they met. Not always easy, but people will respect you if you are very clear about your boundaries and time management.

3. **Step away from your email:** Switch off your work email, not just in the evening, but schedule a time in the morning and afternoon to catch up with messages and turn off your notifications in the meantime. Trust me, this will free up a surprising amount of time.

4. **Close down your work at the end of the day:** Whether you are working from home or in an office, make sure there is a clear end to your working day. Make this obvious such as shutting down your computer, cleaning your equipment ready for the next day, or perhaps writing out a list of things to do first thing in the morning when you start work again. Then, leave it and preferably insert a gap—a fertile void—between this moment and your home life, as we have seen earlier.

5. **Dial down perfectionism:** Striving to be perfect undermines productivity, and well-being, for many people. While we all want to do a good job, there are times when we have to accept that something is good enough for the time and energy we have available. I once did an event with a journalist who was asked how he coped with people saying they could have done a better job of writing his column than he did. "They probably could," he said, "but can they do this every single week, month after month, year after year?"

Change or challenge your perspective

My favorite place in Paris is the Musée d'Orsay. It is packed full of Impressionist masterpieces that have lost none of their allure despite being reproduced on a million postcards. Of all the paintings, my favorite is Monet's *Régates à Argenteuil*. As you enter the room housing this remarkable image, the shimmering light of sunshine dancing on the Seine catches your eye and the water seems to ripple. However, as you move closer to the painting the strokes of color seem to fragment and the impression of movement of light on water slips away. It's only when you stand back that the magical scene appears again.

This simple experience shows us how much our perspective on things depends on where we are standing. This is also true for the complex decisions we make in our personal and business lives.

Sometimes shifting our perspective on things, even slightly, can reveal a new impression and open our minds to different possibilities.

How to transform negotiations

In October 1962, Soviet Premier Nikita Khrushchev sent two letters to the US President John F. Kennedy. These letters were written at the height of the Cuban Missile Crisis. The two letters had very opposing tones: the first was quite conciliatory but the second insisted that the Soviets would not remove their nuclear weapons from Cuba unless the US withdrew their own nuclear weapons from Turkey. In the Oval Office with his advisers, President Kennedy concluded that there were only two options available: either withdraw US weapons from Turkey in return for a withdrawal of the Russian weapons, or initiate a nuclear strike against the Soviet Union within days. To everyone's surprise, though, Llewellyn "Tommy" Thompson, a senior adviser who was normally reticent in such discussions, piped up: "I don't agree, Mr. President." Instead, he advised the president to respond to the first, more conciliatory, letter. He was convinced that Khrushchev could be persuaded to withdraw their weapons if the Soviet Union could be seen as having "saved Cuba."

Thompson certainly understood the Soviet mindset well—he had served as America's ambassador to the Soviet Union and had developed a unique and personal friendship with Nikita Khrushchev during his time in Moscow. US Secretary of State Dean Rusk described Thompson as "our Russian in the room." So, on Thompson's advice, Kennedy offered that if the Soviet weapons were withdrawn, the US would pledge to never invade Cuba. And Khrushchev conceded. This deal allowed him to declare that he had saved Cuba from attack and therefore fulfilled his own core interests of consolidating his power and saving face.

Thompson's ability to see the situation from Khrushchev's perspective gave the US a powerful advantage in negotiations and helped to save the world that day. This ability to change your perspective to match the situation is a fundamental component—the C—of agility. Having an intimate knowledge and understanding of the fundamental interests of your opponents is essential to allow you to challenge and change your perspective in a flexible way to help you achieve a successful outcome in any business, political, or personal negotiation.

Perspective taking versus empathy

The ability to consider the world from another's viewpoint is known as *perspective taking*. It is often conflated with empathy, but they are not the same thing. Empathy is genuinely feeling someone's pain or emotional reaction. Perspective taking, in contrast, is the ability to see someone's standpoint through the lens of your own interests. This capacity allows you to step outside your own biased frame of reference and develop a more balanced perception of fairness in competitive contexts.

Empathy is important, of course, and is often the door through which we learn to build perspective taking. I've little doubt that Tommy Thompson had a lot of empathy for Khrushchev, and this enabled him to see the problem from the Russian's viewpoint. But Thompson did not let empathy cloud his judgment. Too much empathy can lead to preferential treatment of others at the expense of your own self-interest and can be detrimental to closing a deal in competitive negotiations. Therefore, perspective taking is more important for successful negotiating skills than empathy.

Can you take another person's perspective?

To find out where you sit on the perspective-taking and empathy dials, answer the following questions. For each item, think carefully about how it describes you and choose the appropriate number on the following scale:

0. This is not like me at all
1. I'm occasionally like this
2. This is like me some of the time
3. I'm often like this
4. This is like me almost all the time

1. When things don't work out for someone, it doesn't really bother me.
2. I know I've been fortunate in life, and I feel compassion for those who aren't.
3. I try to take everyone's point of view into account before coming to a decision.
4. When other people have problems, I don't get overly concerned.
5. To better understand how a friend might be feeling, I try to imagine how they might see things.
6. I often get quite protective of people who are vulnerable.
7. It's very difficult for me to see things from another person's perspective.
8. When I believe that I am right, I think that listening to others is just a waste of time.
9. I have lots of sympathy when someone is a victim of prejudice.
10. I'm a very gentle and compassionate person.
11. I become overwhelmed and emotional by things I see.
12. I always try to see every side of an argument.
13. When I have to give someone negative feedback or criticism, I try to imagine how they will feel before telling them.
14. When someone offends me, I try to see the situation from their point of view.

How to Get Your Score:

You will have a score of 0 to 4 on each of these questions. The first thing you need to do is to *reverse* your score for questions 1, 4, 7, and 8. So, if you scored 4, give yourself a 0, if you scored 3, give yourself a 1, and so on as below:

4 = 0
3 = 1
2 = 2
1 = 3
0 = 4

Now split the questions into two groups of seven.
Perspective Taking: Add together your scores for Q1, Q3, Q6, Q8, Q11, Q13, and Q14.

Empathy: Add together your scores for Q2, Q4, Q5, Q7, Q9, Q10, and Q12.

This will give you a total score between 0 and 28 for each dimension.

A score between 0 and 9 = Low
A score between 10 and 18 = Medium
A score between 19 and 28 = High

Most people score medium on each of these scales, but it can be interesting to see whether your score differs across empathy and perspective taking. If you have a low score on either one, this gives you an indication of what you might want to work on. No score is especially good or bad in itself, it depends very much on the context. But knowing if you are high or low on either dimension can alert you to potential dangers in particular situations. If you are very high on empathy, for instance, you might want to tone this down a bit if you are in a tough negotiation and focus on trying to turn up your perspective-taking dial. On the other hand, if a friend has just had some bad news you might want to turn up your empathy.

How to change your perspective

Changing your perspective and looking at things from different points of view can transform your ways of thinking about problems and decisions. The reason why we often get stuck in life is because we tend to assess our situation from a single perspective. If you develop the habit of looking at situations from different angles this will really help you to become more agile in your thinking and your problem-solving. This will allow you to create a space to consider new ideas and possibilities. Try the following tips to help you shift your perspective:

1. **Visualize a problem from different perspectives:** Look at a problem you currently have, perhaps an important decision to be made at work or in your personal life, and try to think of four alternative solutions to the problem. Let's imagine that you start a new job and quickly realize that it was not what you expected. Rather than complaining and getting angry about your decision, ask yourself what is good about the situation. Are there any unexpected benefits? Can you change anything about the current post to make it work for you? Is there anything you can negotiate? Can you adapt to see the positives of the role?

2. **Sometimes changing your question is all that's needed:** The way you ask your question frames the direction in which you will look for an answer. So, rather than asking, "How can I stop overworking and getting stressed?" ask yourself, "How can I find more time for leisure?"

3. **Nurture your optimism:** Our natural tendency toward optimism is something to lean into. It may surprise you to know that most of us are optimistic about our own lives. Even during the coronavirus pandemic, for example, one study in the US found that people were optimistic about their own chances of not catching the virus, while simultaneously pessimistic about

the chances that others would become ill. Remember that optimism is not necessarily unrealistic. Those who achieve a healthy balance of what I call "optimistic realism" see the future through rose-tinted glasses while knowing that there will be lots of disappointments and failures along the way. Optimists don't ignore the stresses of life, they just approach adversity in more productive ways. When something goes wrong, rather than blaming themselves and seeing it as permanent, optimists look for ways to learn from the setback. We can all change our perspectives to be more optimistic. Try to look for the positives in any difficult situation, surround yourself with positive, rather than negative, material and people. We all know that negativity tends to pulse through social media, so steer clear if possible. It's not sensible to avoid the news completely, of course, but try to limit your time "doom scrolling," which can take a toll on your well-being. Also, think about the people with whom you are spending most of your time. Are they energizing you or are they draining your vitality? Connecting with people who have a positive energy will nurture your own optimism.

4. **Practice mindfulness:** Finding a way to break out of certain thought cycles and into the present moment can be helpful in seeing things from different perspectives. On a regular basis—at least once a day—try to bring your attention to the present moment. Perhaps focus on your breath for a few moments, or really notice how something tastes and smells, or listen to what's happening around you. Make these exercises a habit, as they are also a great way of boosting your agility.

5. **Read fiction:** Reading fiction is also a great way to shift your perspective. Reading gives you the opportunity to engage with many different lives. By empathizing with characters and immersing yourself in their world you can begin to live their life as they see it. Fictional characters allow us to imagine what it is really like to walk in another person's shoes. This boosts

our capacity for both empathy and perspective taking. And science backs this up. Studies have shown that generating aspects of a novel internally, such as what the characters and their surroundings might look like, can make all the difference. It has been called the "mind's flight simulator"; just as pilots can improve their flight skills without leaving the ground, people can enhance their perspective-taking skills by reading an engaging story.

6. **Ask what other people would do:** You can also try a simple game, which I designed as a fun way to imagine how someone else might try to solve your problem. Allocate the names of six people you admire a number from one to six—these could be people you know, or famous people, or even fictional characters. Then roll a die, and spend the next hour getting into the head of the person it lands on, seeing the world as they would. What would Harry Potter do? How would your best friend approach the issue? How would Michelle Obama deal with the problem? Practiced regularly, this gets you used to seeing things from quite different viewpoints.

Develop your mental competence

In February 2003, over four hundred people attended a concert by the rock band Great White at the Station nightclub in West Warwick, Rhode Island. The opening song was accompanied by a spectacular pyrotechnic display. For several minutes, nobody realized the fire was out of control as flammable foam above the stage was set alight, and the fire moved across the top and around the sides of the stage. Everyone thought it was part of the show.

Within minutes the now fast-moving flames engulfed the entire hall with intense black smoke. The nightclub was overcrowded, and the venue and the band manager were found guilty of many offenses.

In addition, the way in which people reacted in the ensuing panic gives us a vital lesson about the importance of our mental processing, especially when we are under pressure. Hundreds of people tried to escape the fire by exiting the club through the front entrance, which was the same way they had come in. This was a devastating mistake. Tragically, three other entrances were completely clear but most people, in their panic, did not consider turning around. Most people relentlessly pushed toward the front in what was a catastrophic mistake. They followed the herd, which is our natural tendency, rather than working out the best way out. The ensuing stampede led to a crush in the narrow passageway toward the entrance, quickly blocking the doorway and leading to many deaths and injuries.

Sadly, the heartbreaking case of the Station nightclub fire is not unusual. People who perish in crash landings or fires frequently do so because they try to escape through the same entrance they used when they entered.

Our mental competence comes down to the proper functioning of what psychologists call "executive functions." These are basic cognitive survival skills that help us to assess and react to any situation and work out what to do if something goes wrong—to find an escape route in this case.

What are executive functions?

Executive functions underpin our mental competence and they are vital for performance under pressure. I have worked for many years on how these crucial mental components can be identified and trained in elite athletes. They are also, of course, critical for everyday decision-making and they are made up of three critical elements:

- **Inhibitory control:** This is a catchall term that includes the ability to resist acting on impulse or habit as well as suppressing irrelevant information so that we can focus on what's relevant. In the case of the fire, the internal pressure is to act on habit or impulse and move toward where you came in, but this tendency needs to be suppressed to figure out whether there is a better solution. Inhibitory control is vital in sport. In a fast-moving passage of soccer play, for instance, a midfielder might need to keep his focus on any number of things at the same time—the trajectory of the ball, the movement of strikers, the movement of defenders, the position of the goalkeeper, possible gaps into which he might pass the ball—all the while suppressing his attention to other aspects of the game that are not immediately relevant. Obviously, this is a dynamic process, as what is relevant changes from moment to moment.

- **Working memory:** This is the capacity to hold information in your head while continuously updating that information. For instance, listening to a story and keeping track of different events relies on working memory. It is especially important in helping us to make decisions while under pressure. To return to the soccer example, this might be when instructions from the coach need to be translated into game plans, new information needs to be held in mind and added to the mix, and alternatives need to be considered, especially when under pressure. Working memory and inhibitory control often work hand in hand, of course; the memory element updates the most relevant information while inhibitory processes suppress what's not immediately relevant.

- **Cognitive flexibility:** We've discussed cognitive flexibility at length in the previous chapter. As should now be clear, it's vital to stay flexible in order to adjust to new circumstances and take advantage of sudden, unexpected opportunities. Cognitive flexibility skills are those brain-based executive functions that facilitate this ability to be agile. As an example, what do you do when your team is suddenly down to ten players, or the game goes into added time and there's an unexpected ten minutes to play, or the other team loses a player? It is the capacity to adjust rapidly to the new circumstances—the opposite of mental rigidity, which happens when game plans are stuck to blindly even when the situation has changed.

Executive functions are vital for mental agility. Developing all three of your executive functions—inhibitory control, working memory, cognitive flexibility—is essential for planning your time, focusing your attention, regulating your impulses, remembering what you must do later in the day, and juggling multiple tasks successfully. They are also, of course, one of the cornerstones of the first pillar of switch craft: *agility*. Along with Adapting, Balancing your goals and plans, Changing your perspective, the D refers to Developing your executive functions, or what we can call our "mental competence."

These fundamental mental skills—executive functions—are vital for fluid functioning in many situations all throughout our lives. In fact, studies with children have shown that when predicting success in later life, the possession of good executive functioning skills turns out to be far more important than general intelligence or socioeconomic background.

Just as a business might invest in basic infrastructures so that they can respond quickly in the event of changing demands, so investing in the psychological infrastructure that's needed for switch craft— your executive functions—comes with a price. Time and energy are required to develop your mental resources—your executive functions—sufficiently to facilitate your mental agility.

Forward planning is one way to support your executive functions. After the fire in the Station nightclub, the Texas State Fire Marshal's Office created the "Have an Exit Strategy" initiative. This encourages people to always have at least two exit routes from any bar or club, recognizing that the best way out may not be the way you came into the building. This is an enlightened acknowledgment of our tendency to operate on autopilot.

How can you develop your executive functions?

We looked at practicing your cognitive flexibility in the previous chapter. My research team has also looked at whether working memory and inhibitory control can be improved by using simple computer games. Brain-training games have had a bad rap in the media for not living up to the hype, and there is good reason for this caution. While there is often great improvement in performance in the games themselves following lots of practice—little surprise there—most people do not seem to improve much on anything else. There is none of what we psychologists call "far transfer" to other life situations.

Nevertheless, working memory and inhibitory control are important capacities that help us deal with complex and stressful situations. The better these executive functions, the more ready we will be to respond agilely to any situation. One benefit is in helping us to control unwanted distractions. So, by improving working memory, we figured, we should be able to help people to get unstuck from the repetitive negative chatter that occurs when we are worried and stressed.

My research team designed a simple computer game to try to improve working memory as well as inhibitory control. The game we used involves a series of letters presented at different places on a three-by-three matrix, one after the other. The letter K might be presented in the upper-left corner of the screen for a few seconds, followed by the letter B in the lower-left corner, followed by the letter P in the right-middle location and so on. People must remember the location of the *previous* letter. So, in the example above—when you see B in the lower-left corner, you must indicate where the K was. Then, when you see the P, you must indicate where the B was. Once people get the hang of this, the game gets harder: now people must indicate the location of the letter *before* the previous letter, so two letters back. Once people have mastered that, it then asks you to remember *three* letters back. This keeps going until people start making lots of mistakes. After several hours of practice, people participating in the study can typically manage up to three or four items back, with some working-memory superstars managing five!

Was it possible, we wondered, that a training exercise like this might help high worriers to manage their distressing thoughts? We recruited large groups of people who self-identified as "high worriers" and asked them to play the working-memory game for about forty minutes every day for at least eight weeks. After weeks of training at home, everyone came back into the lab to see whether there was any improvement. The results were encouraging. We found that bigger improvements in working memory were related to better control of worries in our high worriers. While the improvements were not life-changing, those who had managed to improve their working memory also managed to get unstuck from their repetitive worry and become just a little more agile.

We have also used computer games to try to tackle impulse control problems. In one study, we selected volunteers who were binge eaters along with those who had no problem controlling their eating patterns. We asked each of our volunteers to do one of two things in response to various pictures. If a picture is surrounded by a green bar, they must press a button—a go response as we call it—while if the image is surrounded by a red bar, they should withhold their response; this is called a no-go trial. The trick is that images of sugary and fatty treats—chocolate cake, crisps, and so on—are always "no-go" trials. The idea is to train the brain to resist moving toward enticing treats and help our binge eaters to gain better control over their impulses. When a healthy salad or piece of fruit is presented, it is always accompanied by a green bar encouraging a "go" response. If binge eaters do this on a consistent basis for several weeks, their habit of eating unhealthy foods without thinking about it begins to be undermined and they gradually learn to exert better control over their eating habits.

How can we boost executive functioning in everyday life?

Some everyday exercises can also enhance these same mental skills. Try putting an enticing treat—a piece of chocolate, say—in your fridge, surrounded by healthier options. Then tell yourself that you can have the treat on a certain day, or at a certain time. Every time you open your fridge, you can look at the treat, perhaps even pick it up and smell it, but then put it aside and select a healthier option. Done regularly, this will bolster your mental control skills, helping you to stop acting on impulse. There may be a few setbacks! But over time you will learn how to gain better control over your impulses.

Participating in sport, learning how to play a musical instrument, or learning a foreign language can also directly target some of the key mental resources that are needed to help you stay in control in a crisis. One study in the Netherlands, for instance, followed a large group of young children for two and a half years, starting when the children were six years old. Two groups of these children received regular music lessons, one group received visual art lessons, and a control group had no arts classes. Those children learning music had much larger improvements in their executive functions relative to the other children. This study supports the elusive "far transfer" from music education to general academic success.

To enhance the first pillar of switch craft—*agility*—it is essential to bring together the four different components—the ABCD. Working on each one separately is fine, but it is only when all four come together and we are adaptable, balanced, able to shift our perspective, and mentally ready for anything that we can become truly agile.

Chapter Summary

- Mental agility is made up of four key components (the ABCD of agility): **A**dapt to changing demands, **B**alance competing desires and goals, **C**hange or challenge your perspective, and **D**evelop your mental competence.
- Being open to adapting to new situations is vital.
- Finding ways to balance your life and making sure that how you spend your time is in line with your primary goals is essential to build your agility.
- Finding ways to challenge your perspective is important to open your mind and lay the groundwork for a more agile mindset. Interrogating your beliefs and reading novels are just two ways to help you to see things from different perspectives.
- Mental agility takes energy and therefore building up your mental resources of inhibitory control, working memory, and cognitive flexibility are essential to allow you to perform at your best.

THE SECOND PILLAR OF SWITCH CRAFT

Self-Awareness

CHAPTER 8

KNOW THYSELF

The Ancient Greek aphorism "know thyself" is the first of the Delphic maxims inscribed on a column in the Temple of Apollo at Delphi (the other two are "nothing in excess" and "surety brings ruin"). It is wisdom as old as the hills—or at least as old as the fourth century BCE to which the temple dates.

Such is its import and standing that nearly every great philosopher has touched upon its meaning throughout the ages. In Plato's *Phaedrus*, Socrates employs the term "know thyself" to explain why he is indisposed to rationally explain mythology or other intellectual endeavors. "I am not yet able, as the Delphic inscription has it, to know myself; so it seems to me ridiculous, when I do not yet know that, to investigate irrelevant things."

Samuel Taylor Coleridge, in his poem "Self Knowledge," refers to the maxim as "the prime and heaven-sprung adage of the olden time!" And Benjamin Franklin, in his *Poor Richard's Almanack*, acknowledges its difficulty thus: "There are three things extremely hard: Steel, a diamond, and to know one's self."

In fact, the maxim even predates ancient Greece. In his classic military treatise *The Art of War*, written in the fifth century BCE, the Chinese philosopher-general Sun Tzu includes the dictum "知彼知己，百战不殆，" which roughly translates as "know others and know thyself, and you will not be endangered by innumerable battles."

Being self-aware means that we are aware of our own thoughts, emotions, and actions and how they might impact on other people. When we are highly self-aware, we are better able to reflect on how we can do better and what aspects of our behavior we might need to modify. This means that the more self-aware we are, the more agile we can be. This is why self-awareness is the second pillar of switch craft; because it is only by understanding our values, goals, and capabilities that we can really respond agilely to any situation. In this way, the second pillar of switch craft—*self-awareness*, or "knowing ourselves"—supercharges the first pillar, *agility*.

But how, precisely, do we go about "knowing ourselves"? And is it really as difficult as Benjamin Franklin feared?

Fortunately, modern-day science can provide an answer to both of these questions—and it's good news on both fronts. We can "know ourselves" physiologically through familiarizing ourselves with and carefully monitoring our internal bodily states. And we can know ourselves psychologically through accurate assessment of our unique personality style. Our personality reflects our fundamental habits of how we think, how we feel, and how we do things in different situations. Unlike temporary mood states, personality traits tend to be consistent over time and in different contexts.

What's my type?

Let's start with personality. My husband's father was a cheeky-chappie London market trader, a little bit like Del Boy from the long-running British sitcom *Only Fools and Horses*. He used to come up with all sorts of funny sayings but one that particularly sticks with me is: "You can always tell what someone is like from their personality!" When we meet a person for the first time, we often get a sense of who they are in terms of their characteristics. Are they extraverted? Do they seem open-minded? Are they conscientious? The idea that each of us has a few enduring personality traits has been common throughout history, and if social media quizzes are anything to go by, many of us are fascinated by the question: What type of person am I? What we are really asking with this question is, "What is my personality?"

While there are many ways to think about personality, it is generally understood to be those individualities that give us a sense of how someone is likely to feel, think, and act under different circumstances. Think of some of the people you know. How would they respond in a minor car crash? How would they react if they lost their job? What would they do if they won the lottery? I bet you can make a fairly good guess based on their personality. People do surprise us, of course, but generally they behave in fairly consistent ways. Our brain craves this type of understanding of others because it reassures us that we know how a friend, a colleague, or a stranger is likely to react in different situations. This understanding has been called "the psychology of the stranger" and it refers to the understanding of a person, or indeed even yourself, at the level of personality traits. It's called the psychology of the "stranger" because personality traits do not tell us necessarily about a person's core beliefs and values; as we will explore in the next chapter, it tells us more about consistencies in habits of thinking, feeling, and acting.

How can we measure a person's personality?

So, what is the best way to capture the uniqueness of our personality? Identifying the key facets of personality has a long history in psychology and there have been many attempts to categorize people into different "types," the most famous of which is Myers-Briggs. Millions have been earned by recruitment firms grouping people into different "types" based on the Myers-Briggs indicator, with a remarkable 80 percent of Fortune 500 companies apparently using the quiz to assign people to the right jobs. The test attempts to help people understand their tendencies based on four "types": *Introvert* versus *Extravert, Intuitive* versus *Sensory, Thinking* versus *Feeling, Judging* versus *Perceiving*.

Many people don't realize that the Myers-Briggs test was developed more than seventy years ago, well before psychology became an empirical science, by Katharine Briggs, a teacher, and her daughter, Isabel Briggs Myers, a fiction writer—neither of whom had any formal training in psychology. The test has many problems and psychologists tend to be highly skeptical about its use. In fact, a meme that psychologists often share on social media is titled "Myers-Briggs is astrology for those with LinkedIn profiles." For one thing, if you take the test on several occasions you will likely get different answers, so it is not considered to be a reliable assessment of enduring personality traits. But even more problematic is that the test tries to divide people into stark black-and-white categories such as "thinker *or* feeler," when the reality is that each of us has different degrees of all of these dimensions. By pigeonholing people in this way we actually separate out people who are much more similar to each other than they are different. And yet, people love taking this test and finding out their "type," perhaps because it is a helpful starting point for self-exploration—even though psychologists don't hold much store by it.

Personality traits are on a spectrum

The more recent consensus from decades of scientific research is that human personality is best mapped onto a spectrum rather than separated into types. What has been found is that five broad dimensions or "traits" (often known as the "Big Five") capture the totality of human personality. Each of these dimensions is measured on a spectrum from low to high: *Openness to experience, Conscientiousness, Extraversion, Agreeableness,* and *Neuroticism.* (The acronym OCEAN is a useful aide-mémoire here.)

To get an indication of how you fare on each of these core elements, or personality traits, try answering the questions below. For each question give yourself a score from 1 to 7 using the following scale:

1 = Disagree strongly
2 = Disagree moderately
3 = Disagree a little
4 = Neither agree nor disagree
5 = Agree a little
6 = Agree moderately
7 = Agree strongly

I see myself as:

1. Extraverted, enthusiastic
2. Critical, quarrelsome
3. Dependable, self-disciplined
4. Anxious, easily upset
5. Open to new experiences, complex
6. Reserved, quiet
7. Sympathetic, warm
8. Disorganized, careless
9. Calm, emotionally stable
10. Conventional, uncreative

How to Get Your Score:
You will have a score of 1 to 7 on each of these questions. First, work out the *reverse* score for each of the following questions: Q2, Q6, Q8, Q9, and Q10. So, if you scored 7, give yourself a 1, if you scored 6, give yourself a 2, and so on as below:

7 = 1
6 = 2
5 = 3
4 = 4
3 = 5
2 = 6
1 = 7

Now, work out your score for each dimension:

Openness to experience: Score for Q5 plus score for Q10

Conscientiousness: Score for Q3 plus score for Q8

Extraversion: Score for Q1 plus score for Q6

Agreeableness: Score for Q2 plus score for Q7

Neuroticism: Score for Q4 plus score for Q9

For each trait, you should have a score between 2 and 14. The overall assessment for each trait is below:

A score between 2 and 6 = Low

A score between 7 and 10 = Medium

A score between 11 and 14 = High

The labels are fairly self-explanatory. Your score on *Openness to experience*, for instance, refers to the depth and complexity of your mental life and experiences, and typically reflects your willingness to try out new things and explore new places and ideas. *Conscientiousness* reflects your tendency to be diligent and hardworking and your desire to do a job well. This trait overlaps closely with your degree of grittiness and persistence. *Extraversion* reveals the extent to which you enjoy being sociable and outgoing. If you are introverted—with a score at the low end of this spectrum—you are more likely to recharge your energy primarily from solitude rather than from other people. *Agreeableness* is the degree to which you are concerned with being "nice" and not offending others; and *Neuroticism* is the degree to which you are prone to anxiety, worry, low self-esteem, and depression.

Understanding your personality is about probabilities and tendencies, rather than black-and-white certainties. While your personality traits do not provide you with the rich backstory that your experience of life provides, this level of understanding is nevertheless important in the quest to deepen your self-awareness. Knowing how your typical traits, or what I prefer to call personality "styles," can influence how you are likely to react in different situations is useful. If, for example, you are introverted, scoring very low on extraversion, you are unlikely to thrive in a highly stimulating environment. You will derive more energy from solitude or socializing with a small group of friends—perhaps an intimate dinner party—rather than a raucous party.

The personality trait that is most important for switch craft is openness to experience. If you score low on openness to experience, you are likely to prefer routine and feel particularly uncomfortable with uncertainty. Holding on to cherished beliefs will be a source of great comfort to you and because of this you may be particularly resistant to change—you might even be at risk of mental arthritis. If this sounds like you, you can take small steps to become more open. For instance, maybe you can start by questioning authority figures. Rather than always accepting the status quo, ask yourself whether there might be an alternative way of doing things or way of thinking. I know it is not easy, but taking the first small steps helps. Try to become more excited about trying out new sensations and ideas. It helps to look at others as role models and try to emulate them, at least to a small extent. More open people typically have a wide range of interests, are usually very adaptable and intellectually curious, and often easily bored. They tend to be introspective and interested in exploring their inner and outer worlds, and are often creative and comfortable with uncertainty and capable of behaving in quite unconventional ways. Remember that there are no "right" or "wrong" traits, but the more open you are the easier you will find adapting to change.

It's important to remember, however, that these traits can be modified and changed—they are not set in stone. While we all have a strong preference, which is indicated by our personality traits, we can learn to modify our fundamental preferences when the occasion calls for it. I am a natural introvert, for example, but I have learned to be quite extraverted when I am giving public lectures or talks at festivals.

Intellectual humility

Scoring relatively high on openness to experience means that you enjoy seeking out new experiences. An often-forgotten aspect of openness is the capacity to accept that your beliefs and opinions may be wrong, and it comes with a willingness to accept that changing your mind is sometimes the right thing to do. This willingness to reconsider your views is called "intellectual humility," and we are only just beginning to realize the important role that this tendency plays in our psychological well-being. Studies have confirmed that those who score high on intellectual humility are indeed more open to other people's points of view and are more willing to consider a range of possibilities on any given issue. This of course is crucial for agility. Most of us tend to overestimate our ability or knowledge in any given area. For example, a 2018 survey found that almost 80 percent of people believed they were more "open-minded" than most people and, perhaps more worrying, fewer than 5 percent of people believed they were "close-minded."

Psychologists have broken intellectual humility down into three key elements:

- Respecting other people's viewpoints
- Being able to separate your ego from your intellect
- Being willing to revise your opinion if new evidence suggests that you got it wrong

Those who are willing to admit that they might be wrong are often happier and healthier than those who refuse to countenance the possibility of being wrong. Given the relative consistency of this habit of mind, it has been suggested that *humility* should be added as a sixth dimension of personality, making a Big Six rather than a Big Five.

Intellectual humility can be nurtured—with difficulty!

Intellectual humility doesn't come to us naturally because many psychological mechanisms, such as cognitive inflexibility, can work against us nurturing a humble way of thinking. Even scientists, who are trained to constantly question everything, are often deeply reluctant to change their own beliefs and abandon theories they have worked on for years. When you have sunk a lot of time and effort into bolstering a particular belief system, or an influential theory, it becomes difficult to acknowledge that you might have got it wrong.

The well-known social psychologist John Bargh conducted a study with some colleagues back in 1996 in which they found that simply reading words that related to elderly people resulted in their young study participants walking more slowly than usual as they left the testing room. Infusing young people's minds with thoughts of old age, it seemed, had slowed their movement. The study became an instant classic. It attracted widespread media attention and sparked the intriguing idea that simply priming people with a stereotype of old age could lead people to behave in a way that is consistent with that stereotype.

Fast-forward to 2012 and a group of Brussels-based psychologists tried to replicate this, now classic, study with a larger number of participants and much more accurate measures of walking speed. They could not reproduce the original results. Instead they found that study participants only walked slower when the *testers*, those conducting the experiment, knew which group of people had been primed with the prompt words, so the testers themselves were predisposed and *expected* the participants to walk slower. The results, concluded the Belgian scientists, seemed to relate more to the minds of the testers than they did to the minds of the study participants. This reflects well-known "expectancy effects" in psychology, which means that when someone knows what should happen, they often leak subtle cues unintentionally to others, causing a self-fulfilling prophecy.

Bargh was furious; he queried the competence of the Belgian scientists, criticized the quality of the journal that published their work, and denounced an article by a leading science reporter on the new study as being "superficial . . . online journalism."

When beliefs that make up part of who we are are challenged, we get upset and tend to double down on our beliefs and become even more resistant to change. Intellectual humility protects us from this powerful psychological mechanism. It is important for switch craft on two fronts: first because it helps us to develop our self-awareness (Pillar 2) and second, it is associated with a greater degree of mental agility (Pillar 1). For example, in a fascinating line of research it has been found that those who are intellectually humble are able to come up with more possible uses of everyday objects in the "Unusual Uses Test," which you might recall is a measure of cognitive flexibility, the base riff of agility.

How to rate your intellectual humility

To give you an idea of how you rate in terms of intellectual humility you can answer the nine questions below. As with the Big Five quiz, for each question give yourself a score from 1 = Disagree strongly to 7 = Agree strongly:

1. No one would ever accuse me of laying down the law, I can accept it when I make a mistake.
2. I really appreciate very clever people.
3. I don't think that changing your mind should be seen as a form of weakness.
4. I appreciate getting feedback from others, even when it's not very complimentary.
5. If I'm ignorant of the facts, I'm willing to hold my hands up and say so.
6. I find it very difficult to laugh at myself.
7. I am open to persuasion by a good argument.
8. When someone criticizes my thinking, I usually feel very uneasy.
9. When someone doesn't get what I am saying, I usually think that they just aren't very bright.

How to Get Your Score:
You will have a score of 1 to 7 on each of these questions.

First, work out the *reverse* score for the following questions: Q6, Q8, and Q9. So, if you scored 7, give yourself a 1, if you scored 6, give yourself a 2, and so on as with earlier questionnaires.

When you add up all of your scores across all nine questions, you should have a score between 9 and 63. Higher scores indicate a higher degree of intellectual humility.

A score between 9 and 21 = Very Low
A score between 22 and 38 = Low
A score between 39 and 50 = Medium
A score between 51 and 57 = High
A score between 58 and 63 = Very High

How to nurture your intellectual humility

There are several ways that we can boost our intellectual humility and these all revolve around finding ways to take other people's opinions on board, to question our own beliefs, and to be open to feedback even if it is painful. No one likes to be wrong, but sometimes accepting that we are can be transformative and helps us learn.

1. **Listen carefully** to viewpoints that you don't agree with without interrupting and do not ridicule the person expressing this viewpoint, even if you don't agree.

2. **Nurture a growth mindset:** Our intellectual humility can be increased by nurturing a growth mindset, which is the sense that our ability is not fixed but can be encouraged and improved through hard work and good strategies. When you are open to learning, you are also more likely to accept that you might sometimes be wrong. It's important to try to avoid believing that your talent for something is fixed and unchangeable— instead keep going until you see some improvement. For example, if you don't feel particularly skilled at playing a musical instrument, keep working at it and you will see improvement in your skill, and you'll also be nurturing a growth mindset by seeing and believing in self-orchestrated change.

3. **Celebrate your failures:** This is easier said than done, I know. But we can only learn from our mistakes if we make them. So, when something does not work out well, have a proper debrief about it. Listen carefully to feedback from as many people as

possible. Ask yourself whether there is anything you could have done to have turned things around. Only by taking all this information on board can you really learn.

Become aware of your intellectual humility and what you might need to work on. Being aware of where you sit on a spectrum of openness and intellectual humility will give you an important insight into a level of self-awareness based on your personality tendencies. This will give you a deeper understanding of yourself and help you to build a stronger second pillar of switch craft.

Knowing your mind and body

Understanding your personality traits is not enough. There is more to "knowing thyself" than the insights that result from psychological self-assessment. To truly get to grips with who you are, you must get to know your body as well as your mind. This is something that the Ancient Greeks were also well aware of. It was the pre-Socratic philosopher Empedocles (c.494–c.434 BCE) who first came up with the notion that there were four "temperaments" composed of the natural elements: air, earth, fire, and water.

Of course, it is Hippocrates (c.460–c.370 BCE) who is credited as first putting forward the theory that the four basic temperaments—sanguine (social, extraverted), choleric (independent, decisive), melancholic (analytical, detail-oriented), and phlegmatic (quiet, easygoing)—were caused by an excess or lack of bodily fluids (or "humors"). The precise combination of these humors was thought to cause either sickness or good health and underpinned all human moods, emotions, and behaviors. These days, psychologists and brain scientists are more likely to attribute personality difference to hormones, neurotransmitters, and other extracellular chemical messengers (rather than to blood, black bile, yellow bile, or phlegm), but a cursory look back at the Big Five personality types described earlier in this chapter reveals just how close Hippocrates was to the conclusions of modern-day theorists.

Although there was a fair amount of time devoted to musing, self-reflection, and pondering the finer points of life in Ancient Greece, it is safe to say that contemporary society is drifting toward sluggishness. Few of us spend our days in physical jobs, with millions sitting at desks in front of computer screens for hours on end, before spending still more hours in front of a screen for our leisure time. Even our domestic chores, like cooking and cleaning, now require less physical skill and effort compared with earlier generations who were not blessed with many of the time-saving gadgets that we now have. I remember my grandmother spending hours crushing fruit in a large bowl by hand to make jam—a task that can now be done in seconds in a powerful blender.

Physical exercise is now something most of us schedule into our busy lives rather than being built into what we already do.

We have become disconnected from our physical reality

This retreat from the physical has disconnected us from our bodies and made us less able to read the faint and diffuse signs of discomfort, pain, and impending fatigue. Your capacity to perceive temperature, itch, tickle, sensual touch, blushing, hunger, thirst, muscle tension, and a host of other bodily signals forms the foundation of your physical self. Understanding these subtle signals from your body is a unique part of who you are. This almost forgotten level of self-awareness—tapping into the physical reality of your own body—is now making a comeback in psychological science.

At the dawn of scientific psychology in the USA in 1884, Harvard psychologist William James developed a theory suggesting that emotions occur as the *result* of bodily reactions to events. He effectively reversed how we usually think about emotions—rather than, for example, running because we are afraid, James believed that *we are afraid because we run*. If you see a snake, your heart does not beat faster because you are scared—seeing the snake increases your heart rate, and when that increase is detected by the brain you begin to feel afraid. James did not present direct evidence for this intriguing idea and it did not gain much traction in psychological circles for many years. Now, with new ways to measure brain and bodily activity, the importance of our bodily awareness to how we feel, think, and act has come back to the forefront of psychology and neuroscience. William James, it turns out, was right.

We now know that the capacity to be aware of what's going on inside our own body, known as "interoception," helps us achieve a more rounded degree of self-awareness. Your level of interoception will reflect to some extent how aroused and energized you feel in a given moment. For example, as you wait anxiously before an important job interview, you may be all too aware of your heart pounding, but when relaxing with friends you are probably completely unaware of your heart beating. But even outside of these contrasting situations, people differ dramatically from one another in terms of how good they are at detecting these internal signals. A high degree of interoception has been linked to heightened anxiety, while at the lower end those who are not very good at tuning in to their bodily signals often have real difficulties in identifying and describing their emotions.

This is an exciting area of contemporary research with many unanswered questions. We still don't know precisely when enhanced interoception can help us and when it can hinder us. Being more aware of a physiological state of fear, for instance, can be very helpful when we are in a threatening situation, but is not particularly helpful when we are about to make an important presentation at work.

Measuring interoception

One of the reasons that bodily sensations were put on the psychological back burner for many years is that measuring internal signals is difficult. They are spontaneous and difficult to predict. One technique is called the "heartbeat detection task" and involves tuning in to your own heart rate. You can try this for yourself. Close your eyes for a moment. Relax. Try to feel yourself breathing. Stay with that for a little while. Now let your awareness shift to less obvious sensations. See whether you can become aware of your heartbeat. This can take some time and you may feel it more strongly in other areas of your body rather than your chest. Once you do find it, try to keep count of the beat. Comparing your estimate of number of heartbeats in thirty seconds or a minute to your actual heart rate can then be used as a rough measure of what we call "interoceptive accuracy."

In general, however, our own insight into our internal sensations is not the best way to measure interoception and researchers are working on improving these techniques. While self-report measures are not ideal either, I do find the following questions can be useful as a brief scan of how good people think they are at reading their bodily sensations.

For each of the questions give yourself 1 for "very rarely," 2 for "occasionally," 3 for "often," and 4 for "almost always," then add these scores together to make a total score, which should be between 10 and 40.

1. I can notice when my stomach feels bloated.
2. When I am startled, I am very aware of what's going on inside my body.
3. I can feel the hairs on the back of my neck standing up when watching a scary movie.
4. I find it easy to notice an unpleasant bodily sensation.

5. I find it easy to pay attention to my breath without being distracted.
6. I am able to focus on a specific part of my body even when there is lots going on around me.
7. I can feel how hard my heart is beating.
8. I am aware of muscle tension in my neck and back.
9. When I take a shower, I am aware of how the water is running over my body.
10. I am aware of having sweaty palms.

A score above 30 indicates that you are well aware of your bodily signals, while a score of 20 or below is a reflection that you may not be overly in tune with your internal signals. A score between 21 and 30 is average.

Interoception and the self

Our body is constantly sending information to the brain about internal regulatory states such as hormone levels, blood pressure, temperature control, digestion and elimination, hunger, and thirst. In fact, we now know that the body actually sends many more signals (80 percent) *to* the brain than the brain sends to the body (20 percent). What this means is that our brain is there to serve our body and not the other way around.

And this setup seems to be important in helping us to differentiate between our self and other people and objects. We know this because our brain sends out a stronger signal to our own heartbeat when we are thinking about ourselves compared to when we are thinking about someone or something else. In other words, thoughts about ourselves pack a bigger physiological punch than thoughts about other people. Studies like this tell us that the mind is best understood as being embedded in a body that itself is embedded in a complex physical, social and cultural environment. Reality is not simply out there to be perceived but, rather, is conjured up in our minds via the constant fluctuations of our own organic matter. Our heartbeat tells us what's most important. This dovetails neatly with the conclusion of the French philosopher Maurice Merleau-Ponty back in 1945 when he wrote: "The body is our general medium for having a world."

Bodily signals help the brain to make its predictions. This point takes on a particular significance when we put it together with the increasingly popular view of thinking of the brain as an inference device that strives constantly to predict what's out there and what is going to happen next. But let's remember that your brain is not controlled completely by what's going on around you. These external signals are combined continuously with signals coming from *inside* your body to generate your perception of the world. It is a highly dynamic, reactive, predictive, and unceasing interchange.

Our bodily signals can also influence what we perceive. This is why a creaking floorboard late at night will seem much more threatening when you are watching a scary movie compared to when you are listening to relaxing music. Your faster heart rate, caused by the movie, enhances the prediction of danger and so the unexplained noise becomes a potential threat signal. The point is that your perception of the world is a lot more influenced by the workings of your internal body than you might like to think.

More importantly, these signals have even been shown to make a difference in the expression of racial prejudice. In the USA, Black people are more than twice as likely to be unarmed when killed during encounters with the police as white people. Potential reasons for this depressing statistic have been explored in laboratory tasks requiring "snap judgments." As an example, volunteers might be presented with photographs or videos flashed up very quickly on a computer screen of people carrying either a gun or a mobile phone. If the person is carrying a gun, the volunteer's task is to "shoot," figuratively speaking, the threatening person as quickly as possible by pressing a "*shoot*" button, and to withhold a response if the person is not carrying a gun. The consistent finding is that white and Asian volunteers are much more likely to make the decision to shoot Black people relative to white people when they are not holding a gun. This is because they are more likely to misidentify an innocuous object in the hand—e.g. a mobile phone or a wallet—as a gun if the object is being held by a Black person.

This knee-jerk reaction seems to be partly informed by what's happening in our bodies, specifically our heartbeat. Further research has shown us that most of the misidentifications occur when Black individuals appeared at the same time as the participant's heartbeat. When the judgment to shoot or not comes in between heartbeats, there is no difference in identifying guns or phones between Black and white targets. When a heartbeat happens, special sensors in the arteries fire off a message alerting the brain. In between heartbeats, these sensors stay quiet. When the brain receives a signal of a heartbeat or an increase in heart rate, it will then generate predictions to figure out what is going on and what needs to be done to stabilize and protect the body. Due to unconscious bias, Black men are often falsely perceived to be larger and more dangerous than equally sized white men. So, the combination of the brain receiving an alerting signal (from the heartbeat) alongside the appearance of a stereotypical threat—a Black man—seems to increase the chances that even something innocuous (a mobile phone in the hand, say) will be perceived as dangerous. In a very real sense, the brain perceives the world through the body. Even racial stereotypes seem to be strongly influenced by the ebb and flow of the body's internal workings.

Perception is an active process

Look closely at the two horizontal lines at the top of the next page. Which one do you think is longer?

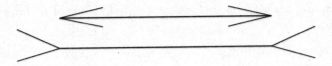

For most people, the lower line looks longer, although both lines are actually the same length. I was captivated by this famous illusion, known as the Müller-Lyer illusion, in one of my early lectures as a psychology undergraduate. It shows us in a powerful way that even our perception of a simple line is not based only on the physics of light entering your eye, but also on your previous experience of the world.

Because most of us live in a very angular world, the brain interprets the arrowheads at either end of the horizontal line as being cues to depth. For instance, the lower figure can easily be interpreted as representing the *inside* corner of a room while the upper line is more like the *outside* corner of a building. As an inside corner the lower line may appear to be nearer and therefore larger. This is because smaller images falling on your retina often represent larger objects that are further away, and your brain takes this into account and recalculates to give you an orderly view of the world. This calculation sometimes leads to dramatic visual errors, such as the Müller-Lyer illusion. There are several studies that show that people who grow up in non-angular environments, such as the Zulu in South Africa and Navajo in North America, who are surrounded primarily by round buildings, are not as susceptible to this visual trick. As this illusion shows, our perception of the world is actively constructed and heavily influenced by our previous experiences.

As with most things, our perception of our bodily signals is also influenced heavily by our experience. Psychologists have long known that the same signal from the outside world can have very different implications for different people based on their prior experiences—as demonstrated by cultural differences in the Müller-Lyer illusion. The same is true of the strong internal signals that alert the brain just as information is coming from the outside world. If you notice that your heart is beating faster than usual when you arrive at a party, say, a more anxious person might interpret that as a sign of impending threat, whereas a less anxious person might interpret exactly the same signal as excitement. These internal signals are almost certainly one of the reasons why anxious people are very resistant to learning that their negative thoughts are irrational, even when the evidence against their belief is strong. The internal signal is conveying threat and is more convincing than other evidence that may conflict with their racing heart.

Sometimes these signals from our body can lead us to make serious errors. As with the racism example earlier, a false stereotype (e.g. that particular ethnic groups are more dangerous) will be activated by our internal signals. However, other gut feelings that tap into our vast reservoir of implicit knowledge can guide us to make better decisions. With these more accurate intuitions, our interoceptive ability can improve decision-making, especially when we are under pressure. When having to make a snap decision, under extreme stress, internal signals such as a heartbeat help us to rely on what we know best. While this may not help sharpshooters (if they hold negative stereotypes about particular ethnic groups) in a crisis situation, it can help us make better decisions when we are well trained for that situation.

An intriguing study with financial traders showed how this might work. There are obviously many factors at play in determining success in the fast-paced world of the trading floor, but one of them seems to be interoceptive ability. A team of researchers studied eighteen traders from a London hedge fund who were engaged in high-frequency trading during a particularly volatile period when there was great uncertainty in the markets. Using the heartbeat detection task we described earlier, traders turned out to be better than average at tuning in to their own gut feelings. This is not overly surprising, as we know that increased stress can lead to improved interoceptive ability, and financial trading is known to induce high degrees of stress. More strikingly, the interoceptive ability of the traders was a good predictor of their overall profitability and how long they survived in the financial markets.

How can we boost our ability to read our bodily signals?

Given the importance of interoceptive ability for self-awareness, it's interesting to ask whether we can improve this capacity. One team of researchers set out to look at the impact of meditation on accuracy in detecting a signal from the body. This is an important study because, while meditation is generally thought to improve an open, nonjudgmental awareness of bodily sensations, there is actually very little hard scientific data to support this suggestion. But this study did show that meditation, or at least body scanning, strengthened the ability of participants to tune in to the sensations of the body.

What is known in mindfulness meditation as a "body scan" is a simple way to release tension that you may not even know you have, and helps you to tune in to your bodily signals. It involves a gradual noticing of each part of your body in a certain sequence, paying attention to any general discomfort, aches, or pains.

- **Get comfortable:** It's best to lie down, if possible, but you can do this sitting in a chair if you are in the office and need a quick stress relief.
- **Take some deep breaths:** Breathe in deeply, letting your stomach expand like a balloon with each inhale. Do this for a few minutes to help you really relax. Perhaps try the "4-7-8" technique, which involves breathing in for a count of four, holding your breath for a count of seven, and then

gently exhaling for a count of eight. This helps you to pause between breaths and really relax.

- **Bring your attention to your feet:** While continuing to breathe gently, slowly begin observing any sensations in your feet. If you notice any tension or aches and pains, simply breathe through it. Imagine the tension leaving your body with each breath. When ready, move on to your calves and your shins.

- **Move your attention through your entire body:** Continue with this body scan, moving upward through your entire body, noticing any pressures, pains, or tensions as you go and continuing to breathe into it. You will finally get to the top of your head—take another three or four deep breaths, visualizing tension leaving your body.

This simple technique can really help you to develop a deeper awareness of your body and is also a great stress buster. If you don't have time for an entire body scan, you can do the same thing in just one part of your body. This is a good routine to practice daily and is especially helpful when you are overwhelmed or stressed.

Developing a sense of your own body and tuning in to your internal physical sensations opens the door to an important level of self-awareness. When you add body awareness to an understanding of your personality traits, you will begin to develop a deeper level of self-awareness and will start to build strong foundations of Pillar 2 of switch craft.

Chapter Summary

- Knowing oneself is a pillar of ancient wisdom common to all cultures.
- Understanding your personality traits is one important level of self-awareness and is also a useful framework to understand strangers.

- Thinking of your personality in terms of "types" is not accurate. Instead, it is more realistic to think of your personality as tendencies that vary along several core dimensions.
- Being open to new experiences and being aware of your intellectual humility is important because these are core aspects of switch craft that can be nurtured and developed.
- Tuning in to your bodily signals, called "interoception," is also important to develop a more rounded sense of self-awareness.
- Internal signals play an important role in shaping your perceptions of the outside world.
- Simple body-scanning techniques can improve your awareness of your internal bodily signals.

CHAPTER 9

BELIEFS AND VALUES

Understanding our personality traits and tuning in to our internal sensations are important elements of developing our self-awareness, as we saw in the previous chapter. However, these aspects only tell us part of who we are. To really understand ourselves it's vital to cultivate a sense of our core beliefs and to uncover our personal narratives—our personal *story*. In this chapter we will explore how you can develop a much deeper and well-rounded sense of yourself, which in turn will help you to understand those around you.

Wednesday, April 27, 1983. 10am. Eleven runners warm up outside the Westfield Shopping Center in Parramatta, Sydney. Among them are some of the world's ultradistance running elite, such as Siggy Bauer, who recently set a 1,000-mile world record running across South Africa. The runners are ready to take on the inaugural ultramarathon from Sydney to Melbourne. They have trained for months and most have corporate sponsors and dedicated support teams to help them cover the 540 miles between the two Australian cities. In their twenties or thirties, these runners are in the shape of their lives.

All, that is, apart from one.

Many of the spectators thought it was a joke when they saw Cliff Young, a sixty-one-year-old local farmer, mingling with the elite athletes. In a press report the previous day a local reporter had warned Cliff that he would not be able to finish the race. Cliff explained that when storms rolled in across his 2,000-acre farm, he regularly had to round up his large flock of sheep on foot, as he could not afford horses or tractors. "It took a long time," he said, "several days, but I'd always catch them. I believe I can run this race."

The spectators, fears were confirmed just after the start. Cliff moved in a strange shuffle and was quickly left behind by the ten other runners who set off at a fast pace. The general wisdom in ultramarathon running in those days was that you should run for approximately eighteen hours and then get at least six hours sleep each night. And so, most of the runners stopped for a rest and some food and a period of sleep after about eighteen hours. But not Cliff. He just kept moving. It was after 2am when he finally stopped for a break and less than two hours later he was off again with his distinctive shuffle style. Unbelievably, at the start of the second day Cliff was in the lead. "I'm just an old tortoise," he told a reporter. "I have to keep going to stay in front."

On hearing this, one of the other competitors, an Englishman named Joe Record, commented: "He says he is a tortoise, but I think the old bastard is a hare in disguise."

Cliff's exploits over the following days caused a public sensation, and five days, fifteen hours, and four minutes after he had left Sydney, thousands of well-wishers lined the streets of Melbourne to cheer Cliff across the finish line. He won the race, finishing almost ten hours ahead of the second placed runner.

In a tribute to Cliff following his death in 2003 at the age of eighty-one, legendary Australian distance runner Ron Grant got it dead right. "Cliff wasn't necessarily the best runner," he told an ABC reporter. "He went out and beat the rest of the field because everyone believed they had to go to bed at night-time, and Cliff didn't read the book where it said you're supposed to sleep." Nowadays, most ultradistance runners get by on very little sleep, just like Cliff.

Beliefs are powerful. The absence of a preconception, as Cliff Young discovered, can leave our minds free to break away and spot opportunities. It can therefore be very useful to look inside ourselves and examine what are known as our core beliefs, which are the source of our many preconceptions of how the world works.

Your core beliefs

It's important to remember that our core beliefs and values are not the same as our personality, which reflects our habitual ways of thinking, feeling, and acting, as we saw in the previous chapter. This is why an understanding of personality traits or types has been called the psychology of the "stranger," because personality traits do not tell us necessarily about a person's core beliefs and values. Our beliefs and values provide us with a much more intimate and individual understanding of who we are. A belief reflects a conviction that something is true, even in the absence of hard proof. Beliefs tend to be highly contextual in that they are based on the cultural and environmental situations we have experienced in the past. A value is a deeper conviction of what's really important and has a powerful influence on how we live our lives. They tend to be less contextual and often reflect universal principles. Values reflect guiding principles such as the importance of integrity, compassion, and so on. Whether we are extraverted, introverted, open-minded, or closed-minded, we still might be guided by the same general values. Our values are important because they can keep us grounded when the world is shifting and changing. While we have to adapt to these continuous changes to thrive, and many of our beliefs may change, our core values need never change.

Uncovering your deepest beliefs

Your core beliefs are at the heart of how you see yourself, other people, and the world at large. They often fly under the radar, lurking within the shadowy reaches of the unconscious mind, but these deep-seated convictions have a profound impact on the way that we feel, think, and act in different situations. Core beliefs often have an all-or-nothing quality to them. Some are very positive (I can do anything once I set my mind to it) but many are self-sabotaging (I am unlovable, I am a failure, other people cannot be trusted).

To uncover a core belief yourself, it is often necessary to sift through many layers of self-talk to get to the bottom of what is underlying everything else. A very effective way to make sense of all those thoughts and self-chatter bubbling away in your head is to keep a thought diary. Jot down an incident that made you feel stressed, angry, confused, or upset (we call this a "critical incident"). Rather than fretting too much about the precise details of the event itself, focus on the thoughts that you had surrounding what happened. There are no hard-and-fast rules in terms of the format, just try to be completely frank and honest with yourself and see if you can get to the bottom of the core belief underlying these thoughts and feelings.

Ask yourself:

1. What happened?
2. How did you feel?
3. What did you do?
4. What were your thoughts and conclusions about what happened?

Let's say your critical incident was something like: *A group of work colleagues went out for a drink after work and didn't invite me.* Now, write down a few words to describe how you felt when this happened and what you did. Perhaps you felt *left out, hurt, upset, lonely, spent an extra hour at work, ignored them the next day.* Now, focus on the thoughts that arose as a result. Maybe you thought: *they find me boring* or *perhaps they simply forgot.* Write out as many thoughts as you can. Now, for each of these thoughts probe a little deeper and ask yourself: What does that mean? Answers might be:

"*Maybe I am not very interesting.*"
    What does that mean?
"*People don't enjoy being with me.*"
    What does that mean?
"*I'll never have a group of close friends.*"
    What does that mean?
"*I'll always be lonely.*"
    What does that mean?
"*I'm a boring person.*"
    What does that mean?

Believing that you are a "boring" person is very black-and-white and sounds very much like a core belief. It is all-encompassing, unconditional, and inflexible.

It's unlikely that your thought diary will unfold as neatly as this, especially at the beginning. Just remember that the goal is to question the meaning of each thought or belief surrounding your upsetting event like a tenacious detective. Think Sherlock Holmes. Eventually you will come to some truth concerning how you feel about yourself.

Another way to help to uncover some of your core beliefs is to simply pose a series of probing questions to yourself. You can come up with your own questions, they just need to try to get at the heart of what you really believe. Here are some suggestions:

Do you think most people are smarter than you?
Do you think that everything you do is wrong?

Is life easier for other people?

Do you think that you can achieve anything if you set your
mind to it?

Are you unlucky?

Are you an interesting friend?

Do you believe that nobody understands you?

Do you feel worthy of love?

Do you think you are attractive?

Are most people good?

Do you tend to use words like "everyone" or "always"?

Regularly reflecting on your thoughts and beliefs in this way will help you to develop a deeper self-awareness.

Your core values

Your beliefs are not the same as your core values

As we mentioned earlier, your personality traits reflect your consistent ways of behaving in different situations, your beliefs represent convictions about the truth or otherwise of certain things, while your values are guiding principles that provide meaning in your life regardless of specific beliefs or personality traits. Your values are, of course, often related to your core beliefs, but they are separate to them and are the fundamental moral codes that guide you through life. So many of us stumble through our daily routines without really thinking about our heart's deepest desires. In fact, it's only when you identify and understand your core values that you can create a life that is rich, full, and meaningful.

So, it is essential to figure out what really matters to you in all aspects of your life. As well as being different to your personality traits and your beliefs, your values are not the same as your goals.

Values are what keep you moving in a certain direction, whereas goals are specific milestones that you want to achieve along the way. You will have multiple goals that will change over time, but your values are an ongoing process.

What are your values?

Below I have adapted a simple exercise to help you uncover your core values from the wonderful book *The Happiness Trap* by Russ Harris. The idea is to try to think of values in each area of your life in terms of general direction of travel—what is really driving you— rather than in terms of specific goals. What's really important to you? What do you really care about?

Reflect on the following themes. For each of the areas of life below, think deeply about what they mean to you and add any that I have left out that are relevant for you. Take some time to decide what sort of person you want to be, establish what is most important to you at a deep level, and identify what you want to stand for in life. Clarifying and being guided by your core values is essential to building a deeper level of self-awareness, the second pillar of switch craft.

1. **Family:** What sort of family member do you want to be? What kind of relationships do you want to build and maintain? Think of each relationship in turn. If you were the ideal child, parent, sibling, auntie/uncle, grandparent, how would you interact with others?

2. **Partner:** Do you want to be in a partnership, intimate relation-ship, or marriage? If so, what sort of partner do you want to be? Ideally, how would you behave?

3. **Work life:** What kind of work would you like to do? What do you value most in your work? If you were the ideal employee, employer, colleague, what sort of relations would you like to build?

4. **Personal growth:** Do you value education and personal development? What appeals to you? What would you like to learn? Be careful not to get caught up in goals here: "I want to learn French" is a goal, whereas "I want to be able to communicate with people in their own language" is a value.

5. **Spirituality:** What is important to you in this area of life? Is being in balance with nature important to you? Is religious faith important? Is being skeptical about a higher presence also important to you?

6. **Community life:** What part do you want to play in your community? Is getting involved in politics or volunteering in community groups important to you?

7. **Self-care:** What sort of person are you? Are you self-compassionate? Is it important to drive yourself hard? How do you want to look after your health and well-being? Is this important to you? Why?

Remember that your values are general principles that give meaning to your life. They are also what keep you grounded and able to keep going when times get tough.

Getting to know your true self

As you move from being a child, to a teenager, to an adult, your beliefs and values become incorporated into a fundamental sense of who you are and where you fit in the greater scheme of life: your "true self." When you ask yourself the critical question, "Who am I?" you must never forget the social and cultural context that has forged you. Your beliefs and values reflect your accumulated biases. Many come from our immediate family and our community, in fact it is typically these beliefs that come to play a disproportionate role in determining who we are. Our political beliefs are among the most difficult to change, and if we are confronted with an opposing view a defense network in our brain fires up to resist working out the implications of a new perspective. This is important for helping to keep our minds open and flexible. As a belief grows stronger and deeper it becomes less and less likely that we will keep our minds open to alternative realities, and more and more likely that we will give priority to information that fits with what we believe. This is "confirmation bias," so we are drawn to information that fits, rather than challenges, our worldview.

What happens when our beliefs are inconsistent with each other?

Because of confirmation bias, among other things, it can be very difficult to change a deeply held belief. To challenge our beliefs means questioning the core of our identity and takes up considerable time and energy. That's why it's so deeply disconcerting for us when we hold beliefs that are inconsistent with each other or with our actions. In a quirk of mind called "cognitive dissonance" in psychology, this refers to the mental conflict we feel when we hold different beliefs that are not consistent with each other. For example, imagine that you strongly believe that fuel emissions are an important contributor to global warming but at the same time you love and frequently drive your aging gas-guzzling car. That conflict, between a belief and an action, sets up an internal motivation to either change your behavior—stop driving your beloved car—or change your belief—you might try to convince yourself that gas emissions from just one car don't really make that much difference.

The mind's imperative is to resolve this tension and reinstate balance. More often than not people don't change their behavior—because that's hard—but neither do they want to challenge a deeply held belief. Our natural tendency is to instead reinterpret the facts in a way that helps to reduce the dissonance. This is why we often continue to hold on to beliefs even when faced with indisputable evidence against them.

The more personal the belief, the less likely it is that we will let it go. Take the case of PJ Howard, who lived in the small town of Ennis, County Clare, in the Republic of Ireland. Life was good for PJ. Over the years he had amassed over €60 million from his successful property business, and in 1998 he fell in love with a beautiful vivacious woman fifteen years his junior, whom he had met in a local store. For eight years, he and Sharon Collins lived together, regularly traveling the world in lavish style. While they never married, PJ and Sharon pledged themselves to each other at a glittering party in Italy that they threw for friends and family in 2005.

Less than a year later, Sharon was arrested for hatching a detailed plan to murder not only PJ, but also his two sons, so that she could inherit his entire wealth. The evidence was beyond doubt. Detailed emails from Sharon to an American-based hit man outlined in great detail her proposal to kill the two sons by making it look like an accident, and then to kill PJ by making it look like a grief-stricken suicide. "Would that be too far-fetched? Is it possible to look like an accident and not a hit?" she asked the would-be assassin in an email. In spite of all this, PJ remained fully committed to Sharon. He simply could not believe that the woman he believed to be deeply in love with him had plotted to have him killed. "It doesn't make sense," he told the court, "I find it very, very, very hard to believe." In his witness statement, he begged the jury not to convict Sharon and, as he left the stand, he kissed her warmly on the lips.

The jury and the police were unmoved, however, and Sharon was found guilty and sentenced to six years in prison. Tragic though it is, PJ's story gives us a perfect example of cognitive dissonance at play. Rather than changing his core belief—that he had a good-hearted partner who was devoted to him and loved him—he simply dismissed any information, no matter how incontrovertible, that conflicted with that deeply held belief. Love, in this case, truly was blind.

Our beliefs provide the basis to understand the world

Our beliefs are the mental scaffolding upon which we build an understanding of the world. They help us to generate a myriad of assumptions and preconceptions that simplify the complexities of our social and emotional lives. We are essentially "cognitive misers" who use beliefs as energy-saving devices to simplify our processing of the world. Imagine if you had to work everything out from scratch every time you found yourself in a new situation. Your brain would quickly become overloaded. So, nature has evolved a clever solution. Rather than computing evidence from first principles, which would take up too much energy and time, beliefs allow us to distil complex information so that we can draw rapid conclusions.

Lots of studies demonstrate that strong beliefs, like stereotypes, free up mental resources, allowing more time and energy to focus on other things. The downside is that accuracy is sacrificed for the sake of efficiency, and we can be lured into making strong assumptions with just the flimsiest of evidence.

I had personal experience of this a number of years ago when my husband Kevin and I were arranging help for his father, John, who suffered with Parkinson's disease. As John's condition worsened, we realized that we could not leave him living on his own. He moved in with us, but the problem remained that he would have to spend long hours alone each day while we were both out at work. We discussed with a local care agency the possibility of hiring someone to help out for a few hours each day. A succession of temporary caregivers then came and went, which was difficult for John as he could never get to know them very well. We worked hard with the local agency to secure a permanent caregiver who could commit to helping John on a more regular basis. We knew it was important to find someone that he could get to know and trust. One morning the care agency called with good news. A young man, Tony, was available. He lived locally and could come in for four days a week over the long term. Great.

When the doorbell rang promptly at 8am on Monday I opened it to find Tony, a heavily tattooed man with a skinhead haircut, five nose rings, and dark glasses, explaining that he was "a bit hungover this morning." My heart sank. Could I leave John in the care of someone who looked so rough and untrustworthy? All my stereotypes kicked in. I'm embarrassed to admit that I even wondered whether he might rob us.

Tony turned out to be the best caregiver possible. He and John bantered about soccer and politics, always laughing and joking. Not only did Tony turn out to be a great cook, making fantastic lunches for John, but he even persuaded him to go out and exercise as much as possible, which previous caregivers had spectacularly failed to do. Tony stayed with us for over a year and became a firm friend and caregiver to John. Thankfully, I was able to overcome my initial alarm and not be driven by my stereotyped thoughts. I'm very glad I did because Tony transformed John's life.

What this experience brought home to me was the critical role our beliefs and preconceptions play in the development of mental rigidity. While they are of great benefit in helping us cut through complexity, they are a bit like wearing blinders. Wearing them, we don't necessarily take in all of the relevant information but instead stick unbendingly to the implications of the beliefs. In my case: *tattoos + skinhead = untrustworthy!* That simple computation, in Tony's case, was completely wrong.

Get into the habit of challenging your beliefs

This is why it is so important to persistently question and query your central beliefs. It's not easy but will help you to open your mind and develop a deeper sense of self-awareness. Not only is this habit important for developing your self-awareness, but it also tackles the possibility that the complex web of beliefs you hold might not actually reflect who you are. Our beliefs also, of course, feed into our values so if we don't fully understand our beliefs we may not be guided by our values in a genuine sense. We are simply going along with the crowd rather than being driven by a deeper sense of meaning. To become truly self-aware, and build the second pillar of switch craft, it is important that how you live your life reflects your own deepest values and not some version imposed by your family, your friends, the part of you that protects your ego, or society at large. If beliefs are the building blocks of self, then your values are reflections of what is most important to you.

The Greek philosopher Aristotle advised us: "Knowing yourself is the beginning of all wisdom." And yet, many of us have little understanding of our core beliefs and our deepest values. As such, we have yet to "find ourselves." While "finding yourself" might sound self-indulgent, I would argue that it is actually an unselfish and important process when done in the right spirit. To be the best parent, the best colleague, the best friend, you first have to know and accept yourself. This is not easy because we often end up hiding our true selves without knowing it. Your brain contains multiple layers of meaning, with echoes of long-forgotten memories and associations that shape your life and lead you to act on automatic pilot. These memories and deep-seated habits can push you to act on society's terms, rather than on your own.

In order to adapt, we often have to be different people in different situations, and this flexibility is a good thing. However, the danger is that we may lose sight of our true self and slip into a situation in which our lifestyle does not reflect our deepest values. For instance, you may be passionate about the value of protecting the planet while working for a large corporation that is not doing as much as it can to protect the environment to keep costs down. To be fully authentic, it's vital to bring together this capacity to be adaptable but also remain true to yourself. There are times when it makes sense to turn a blind eye. But the problem is that much of the time there is tension and conflict between being adaptable and our deepest-held values. We often become reliant on our more superficial selves, rather than recognizing the multitude of other options and choices that are available to us at any given moment. This is why many people conform to social roles that may be at odds with their true selves.

Personal narratives

Stories are us!

Nowadays, what is referred to as our "authentic self," or our "true" self, is a hot topic in psychology. Many studies have asked how we make sense of our true and authentic selves given our multiple responsibilities, interests, and desires. The answer that has emerged is that we can develop a deep level of self-awareness using the stories we tell about ourselves. Our personal stories create meaning for us and become an integral part of who we are. They are the bright (sometimes tattered!) existential ribbons that bind together those unique narrative bundles of deep-seated beliefs and core values that make us who we are. Stories are *us* in a very real sense and can give us access to a deeper, more personal level of our personality.

Revealing your personal narrative

One of the best ways to uncover your true self, then, is to write down a couple of stories or incidents from your own life that capture something important about you and your character. Done honestly, this can be surprisingly insightful. In a series of coaching sessions that I conducted with a small group of business leaders, a participant called Tom had a profound revelation that transformed his life for the better. His stories included an account of how he had jumped into a garden pond to save a baby when he was nine years old. He could vividly remember the fuss that was made of him at the time, and how good he felt. Another of his stories occurred in his early twenties, when he regularly agreed to not drink alcohol so that he could drive his friends home from a night out. Like an intrepid personality prospector looking for nuggets of his true self in the silt of his formative experiences, Tom realized a clear theme was emerging: that he saw himself as a "protector," whose life's narrative was helping people. One of his deep-seated beliefs was that "others needed looking out for" and, accordingly, one of his core values was "safeguarding." This helped him to understand some of his present behavior, and why his wife and children often complained that he was overly controlling.

Our sense of self develops in three stages:

- Actor
- Agent
- Author

When we are very young, we tend to have very clear roles—son/daughter, sibling, friend—and our stories from this time reflect this sense of acting a role in the world. For a six-year-old it's all about the factual details of who you are, what you do, and who does what to each other. As we enter adolescence, we still play these roles, but we also start to develop goals and make decisions that we hope will help us achieve them, becoming agents of our own destiny. Finally, as we enter young adulthood we begin to incorporate our past and present experiences together with ideas about where we want to be in the future. Beliefs, values, and self all merge like paths on a map of identity into what we might call our "narrative identity."

The way that you link memorable life events together and construct a meaningful and coherent story out of them allows your brain to navigate a complex and often confusing world. This is something we all do naturally, and it is important to build our sense of self. Your narrative self is embedded in your personal stories and most of us will share these stories we tell about ourselves with others. It turns out that this is really important. It is by sharing your stories with others that you get the feedback that helps you to refine and expand upon your own self-understanding. You may think that some of the ways you acted when you were younger were unforgivable, but others might tell you that your reactions were completely normal and not unforgivable at all. According to research by developmental psychologists, this is how your memories for events become much more flexible and provide you with opportunities for personal growth.

Organizing our past into a life narrative is a powerful way in which we build a sense of who we are. The stories we tell ourselves matter hugely. The facts of what happens to us in life can be far less important than the way in which we reconstruct those events in our mind. If, for example, you witnessed a gruesome road accident, but have constructed the event into a redemptive narrative of how you developed a deeper appreciation of life as a result, then the event is unlikely to have an enduring negative impact on your well-being. The meaning that you have created to make sense of this event—something that was awful actually led to some positive personal developments—is a sign of psychological well-being. The science tells us that stories of redemption in particular are a strong predictor of psychological wellbeing.

Revealing your life story

Many of us are natural storytellers and you probably already have some narrative stories that you tell regularly. However, some of us don't think much about our life and our narrative stories. Whatever your degree of storytelling, the following exercise can be very useful to reveal some of your central narrative. It may not be a surprise, or it may be a revelation. Either way, it will help you to identify your true self.

How to find your narrative identity

It is important to schedule time for yourself—perhaps an hour—to do this exercise and to find a quiet place where you will not be disturbed. If you are using a computer, make sure that your email and any other automatic notifications are switched off.

In this exercise you are a storyteller, and your task is to draw out the highlights of your own story. Think of your life like a book complete with chapters, key characters, and different scenes and themes. You might want to think about an overall plot summary of your story, going from chapter to chapter. Once you have described the overarching plotline for your story, the task is then to focus on four events that stand out. These don't have to go in chronological order and may not be major events, they just have to be meaningful and authentic for you. So, have a think and choose four key events that illustrate:

1. **A low point in your life:** This is an event that felt very negative and may be associated with terror, disillusionment, feelings of guilt or shame, or utter despair.
2. **A high point in your life:** This is an event that was associated with real joy, happiness, or a sense of satisfaction, relief, or contentment. These are the moments that stand out because they are so utterly positive.
3. **A turning point:** This is a time or an event in which you had a profound change in self-understanding.
4. **A self-defining memory:** This is a memory that reflects an enduring theme in your life. It is usually highly emotional and helps to explain who you are. For example, it was this prompt that Tom was responding to when he uncovered his self-identity as a protector.

Write down a couple of potential episodes for each of the four prompts—this can be instructive in itself—and then choose the one you think is most representative. The "self-defining" memory can be particularly difficult and may only emerge after lots of reflection, so don't fret too much if something does not come easily to mind. Tom's revelation about being a protector, for instance, only came after several months. Once you start thinking about these stories on a regular basis, your key events will gradually emerge. For each of your four stories try to give as much detail as possible about where you were, who, if anyone, was with you, what exactly happened, and how you (and others if relevant) reacted. Try to outline what you were thinking and feeling during the event.

If you want to do this exercise, don't read on for now. To really benefit from this exercise, it is essential to be totally honest with yourself and if you know what researchers look for in these stories, which will become clear in the next few paragraphs, you may be tempted to include them rather than writing from the heart. Come back once you have completed your stories.

Interpreting your life narratives

Welcome back. Hopefully you managed to write out some stories using the prompts above. Now we can look at what you might have revealed about yourself by doing this exercise.

There are three primary themes that often emerge in people's life narratives. Have a look at your own stories and see whether you can identify these themes:

- **Emotional quality:** Is your story broadly positive? Does it start in a bad place, and then end up in a more positive way? Or does it go in the opposite direction—from good to bad?
- **Complexity:** How complex is your story? Does it have lots of rich details or do you just give the bare bones of the event?

- **Meaning-making:** Does your story show that you have tried to draw meaningful lessons from what seem like quite different situations?

The greater the complexity of your story, the more likely you are to make sense out of seemingly distinct events. Likewise, a narrative that has a broadly positive emotional tone, especially if it moves from being negative to positive, is associated with better psychological health. If you want to score your stories in the way that researchers would, you can use the scoring scheme described in Appendix 2. This will give you a score for each of the three primary themes of emotional quality, complexity, and meaning-making, although this is inevitably highly subjective. While some people like this formal scoring, what can actually be most helpful is simply the process of writing out the set of stories. So just reflecting on your stories may help you to see some patterns, as Tom did, and to gain a deeper sense of yourself.

What if your life themes are fairly negative?

Many people find that many of their themes are fairly negative. If this is the case, it is good to have uncovered this—you may not have been aware of just how negative you have been. It may be worth questioning why your stories have a negative tilt—is it just the mood you are in today, or is this a larger habit? There are lots of exercises throughout the book to help you to deal with negative thoughts (see Chapter 11) and how to challenge your perspective (Chapter 7) that you should find helpful.

Your true self and switch craft

Developing self-awareness is important because when unexpected events happen a good awareness of our personal preferences, our beliefs, our ways of doing things, our biases and interpretation of past and present events, and especially our values, all help us to understand why we react the way we do. In this way, self-awareness can help us to step back and respond more effectively, bolstering our ability to be agile and helping us adapt to the new situation.

We can develop a deeper sense of self-awareness by exploring our personality traits, our bodily signals, our core beliefs and values, and those personal narratives that we have incorporated into a sense of who we are.

Chapter Summary

- To build the second pillar of switch craft, self-awareness, it's not just about personality traits, intellectual humility, and bodily awareness—we need a good understanding of our beliefs and values. What you believe in life, about yourself and others, is extremely powerful. Your beliefs can keep your mind closed and blind you to who you really are.
- Your most cherished beliefs are very difficult to change, but you should start by taking the time to identify what your beliefs are and challenging them.
- It's also vital to identify and connect with your core values, the things that are most important to you in life.
- Together, beliefs and values combine to form your "true self."
- You can often find keys to your true self in personal stories that mean a lot to you. Your authentic self will emerge through these stories.
- These personal stories are the everyday narrative crucibles where beliefs, values, and meaning come together as one, and they can help you build greater self-awareness.

THE THIRD PILLAR OF SWITCH CRAFT

Emotional Awareness

CHAPTER 10

UNDERSTANDING YOUR EMOTIONS

On the morning of April 9, 1986, I was staying with my parents in my childhood home on the outskirts of Dublin when I was disturbed by a loud knock on the front door and the sounds of sirens screaming outside. Opening the door, I was startled to see four armed members of the Gardaí—the Irish police.

"Has anyone passed through your back garden?" one of them asked.

"No," I replied, "I don't think so."

"Mind if we take a look?" They proceeded through the house and went out into the garden, where they looked in sheds, behind hedges, and in the garage. Helicopters were circling overhead. Something major had clearly happened.

"What are you looking for?" my mother asked, but they wouldn't tell us.

Sometime later we heard the shocking news that one of our neighbors, Jennifer Guinness, had been kidnapped. Jennifer was the mother of one of my childhood friends, Tania, and I had often spoken to her as a child when we would drop in for tea and toast after visiting the beach. She had been taken by a group of men who burst into her house armed with Uzi machine guns. Like most Dubliners, we became fixated on the TV news reports about her kidnapping. The kidnappers, under the mistaken impression that she was part of the extremely wealthy brewing Guinness family, were demanding a huge ransom payment. Thankfully, she was released unharmed eight days following her capture, and the neighborhood gradually drifted back to normal.

I bumped into Jennifer a short time after her ordeal when I was out walking, and I was curious to hear more. She was a resilient, no-nonsense type of woman, so I wasn't surprised to hear that she had remained calm throughout the eight days, despite fearing for her life. She told me that she had generally been treated well, and had observed her kidnappers closely. One of the older men had been menacing and so she was very wary of him; however, one of the younger men seemed softer and somewhat unsure of himself. She took a strategic gamble, and occasionally became angry with him, shouting and ordering him to let her go. Her instinct was that while displaying anger would have been dangerous with the more intimidating kidnapper, it might be effective in unsettling the younger, less confident, one. We can only speculate as to whether this strategy was effective or not in this particular situation. But I think it likely that it at least helped Jennifer to feel more empowered in what must have been a terrifying situation.

Why emotional awareness is essential for switch craft

Several years ago, my husband Kevin wrote a book about the art—and science—of persuasion. In a quest to uncover the DNA of social influence he did something rather unusual. Not only did he spend time interviewing some of the leading academic experts in the field, he also hung out with some of the world's top confidence artists—not very nice people, but geniuses of persuasion who'd never read a textbook in their lives but who had worked out the tricks of their trade from first principles.

Remarkably, there was a significant convergence between the academic and the "practitioner" groups over what makes a good persuader and what constitutes a powerful persuasive message. Two components of successful influence stood out.

Firstly, the message—what you are saying—must appear to be in the other party's self-interest.

Secondly, the messenger—the person who is doing the persuading—has to appeal and have credibility.

To be a good persuader, in other words, you must:

1. Have complete awareness of your own and others," emotions, and be able to express them consummately like a professional actor on the stage of influence.
2. Be able to select (a) the right emotion; for (b) the right message; in (c) the right context; for (d) the right individual or target audience.

Sound familiar? That's right! To be a good persuader you need to be good at switch craft. Or, to put it in more familiar terms, it's not just what you say, it's how you say it. Understanding your emotions and being able to modify them rapidly is a powerful tool in helping you to adapt—whether in persuasion or other situations.

Anyone who has ever turned on a television knows about the "good cop—bad cop" routine. In real-life interrogation suites and interview rooms such manipulation of emotional expression is a highly effective means of extracting information from those on the other side of the table.

But there are more subtle manifestations of emotional manipulation that we encounter daily. Visual website optimizers—electronic platforms that allow marketers and product managers to best represent their wares online—harness three separate principles of persuasion, each of which taps into fundamental emotional systems, to maximize revenue potential:

1. The principle of *scarcity*, or FOMO—the *fear* of missing out ("While stocks last!" and "There are only two hotel rooms left at this price!" are two great examples).

2. The principle of *reciprocity*—the *contentment* of feeling you have bagged a bargain (whatever your goal as a vendor—encouraging shares on social media, getting potential customers to download a product or sign up for a newsletter—it begins with identifying the best offer you'll give the customer so they will be encouraged to give something back in return).

3. The principle of *social proof—security* in the knowledge that other people just like you have bought into the product or service on offer (most often conveyed by likes, positive reviews, and enthusiastic testimonials).

Next time you venture online to make a purchase, just remember this: you are walking into a switch craft minefield!

So, emotional awareness—Pillar 3—is critical for switch craft. Understanding and manipulating emotions is not only useful for persuading someone to buy something, but also to get yourself out of a potentially dangerous situation, as my neighbor Jennifer had figured out.

Anger can be useful. Emotions can bring about powerful change and Jennifer's instinctive use of anger when she was held captive is backed up by more recent science. For instance, we now know that anger can be a very effective negotiating tool, giving us power in unpredictable situations. We tend to be uncomfortable with anger—and motivated to remedy the cause of it in others. Angry people are also often viewed as more powerful and, in that moment, of a higher status. Angry buyers are more likely to get a better deal on their mobile phone contract, for example, with sellers much less likely to push back on their demands.

If you are thinking about trying out this strategy, however, be careful. Remember that context is crucial. Expressing anger toward those with much more power than you will almost certainly backfire. Either they will ignore you, or retaliate, as anyone who has ever been passed over for promotion and given the boss a "piece of their mind" will attest. However, if your protagonist has lower power than you, then displays of anger can be very effective, especially if it is seen as appropriate. In Jennifer's case, although she technically had low power relative to both her captors, the kidnapper who was more hesitant perceived that Jennifer had the moral high ground and her anger potentially helped to weaken his resolve.

Emotions as levers

Something else we know is that emotions give us a great "reality check." Emotions play a crucial role in helping us adapt to change because they can help us disengage from a cherished goal and move on to another. For example, if our partner has rejected us, we might feel a range of emotions—rage, disbelief, grief, shock, despair— perhaps even relief. Even though they can be unpleasant, emotions such as sadness can help us to switch away from our deeply embedded plans and goals. They help us to enter that "fertile void" before we can begin to move on with our lives. In this way, emotions facilitate meaningful directional change at important moments in our lives.

What this means is that emotions give us great agility in terms of which actions to take to achieve a particular goal. Unlike a reflex that tightly connects a stimulus to an action (moving your hand from a hot surface, say), an emotion allows you to break this link so that when something happens—someone pushes into a queue in front of you, for instance—your emotions allow a variety of reactions. Emotions allow you to observe, gather as much information as you can, and then decide how to act. They are the gearbox, if you like, that links thoughts and judgments to actions. This is why they facilitate agility and are important for switch craft.

Go with the flow

Emotions are the brain's expression of big data—all your experiences, both good and bad, are stored in your brain's memory banks and are called upon to help predict the likely outcome of any situation. Emotions are useful precisely because they help to navigate the critical turning points in your life. To thrive in a dynamic world, it's essential to become more comfortable with setbacks and failures, to embrace them, and to integrate them into a complete, coherent package alongside your triumphs and successes.

The first lesson that science, executive coaching, and psychological therapy all tell us is that there is no substitute for experience. To perform to your best, you must engage with the reality of the everyday. Getting out into the world, savoring the flavors, tastes, sounds, sights, textures, demands, and frustrations that the world offers, provides you with a range of feelings that help navigate complex situations. As the years go on, your quality of life will be determined to a large extent by what you have experienced.

Emotions help us to communicate how we are feeling. Subtle changes in how we feel, or in how other people express their emotions, tell us a lot about what is going on. Expressions of emotion provide clear information to others, which is likely to influence their actions. When you approach a toddler who you don't know and who doesn't know you, for instance, if he looks happy and is smiling directly at you it will almost certainly draw your attention and perhaps stimulate a positive playful interaction. A screaming, scowling two-year-old, on the other hand, might cause you to withdraw for fear of frightening him even more. Emotions are by-products of a sophisticated nervous system and help us to regulate not only our own behavior but also that of others, and this is vital to help us adapt. This is why emotional awareness is an important pillar of switch craft.

To begin our journey to develop a deeper level of emotional

awareness and understanding, it's interesting to explore where emotions come from. By understanding their fundamental nature we can better understand what they're for.

Where do emotions come from?

There are two broad schools of thought in affective science about where emotions come from. Each side has some evidence and there is still no general consensus about which approach is closest to the truth—the jury is still out. These uncertainties are typical in science and require an agile mind to make progress.

What has been called the "classical view" proposes that a small number of the most common emotions are more or less hardwired into our brain. In the 1960s, a popular idea in psychology and neuroscience was that the human brain could be thought of as three separate brains in one (known as the "triune" brain). Each of these three structural regions corresponded to distinctive periods of evolutionary development: The oldest was the ancient *reptilian* core, the base of the brain and top of the spinal cord, which deals with basic functions like maintaining breathing, thirst, heart rate, and blood pressure. The central part of the brain just above that was the *limbic* system, which is tucked in beneath newer cortical material and is the seat of our emotions. And finally the outer layer of the brain, the *cortical brain*, which wraps around the rest of the brain and distinguishes us most from other species, is responsible for implementing constraint and many other higher-level functions, such as language and rationality.

While there is some degree of structural truth to this "triune" framework, the three-brain idea is no longer taken seriously in neuroscience. Nevertheless, the idea sparked an influential field of studies, often with animals, to try to understand the biological nature of emotions. Among other things, these studies revealed the importance of some tiny structures within the central—limbic—brain area that are important for our survival. The most famous of these is the amygdala.

The amygdala—fear central?

The *amygdala* is a tiny structure only about the size of your thumbnail, and is thought to be the brain's alarm system, dampening down activity in all other brain areas when a threat is detected. In evolutionary terms, it's very old: if the brain was a club and had founding members then the amygdala would be one of them. The amygdala has more influence on the cortical brain than the other way around, and this is because many more nerve fibers reach from the amygdala deep into the cortical areas of the brain than go the other way. This mechanical arrangement allows "thinking" to be put on hold while attention is focused on potential danger. This is why we can become frozen with fear if we see a spider in the bath even though our "thinking" brain knows very well that it is harmless. As an alarm system, the amygdala acts fast.

Several years ago, I was enjoying a jog in the warm sunshine around the small Cambridgeshire village where I lived at the time. Suddenly, a large snarling Doberman bolted straight toward me down a long driveway. I could hear the owner shouting, but it was clear the dog was taking little notice. The dog gave chase and snapped viciously at my legs. For the next few seconds, I think I could have beaten Usain Bolt as I sprinted up the road. After about 20 meters, the dog thankfully returned meekly to his owner.

I kept running until I reached the top of the road, where I had to stop and try to calm my pounding heart. I began to tremble uncontrollably and it took about ten minutes before I was able to continue with my run. For months afterward I crossed over the road every time I ran past that house. I never saw the dog again and even years later, after I knew that the family and their dog had moved and were no longer there, I still felt a twinge of apprehension any time I passed by that driveway.

Most of us have experienced something like this. When you have been threatened your body responds in a distinctive way that you cannot control; you feel *fear*. Fear is often considered to be a perfect example of a hardwired "basic" emotion. These are emotions that can be recognized in other species such as primates, rats and mice, and even insects and spiders. According to the classical view of emotion, a set of fundamental feelings such as fear, disgust, anger, happiness, sadness, and surprise each has their own neural circuit, or "fingerprint," and this has helped our ancestors survive across millennia.

The idea that some of our most common emotions are hardwired makes sense and has held sway in affective science for many years. The problem is that this view might be completely wrong. There is an intrinsic interconnectivity among many different brain regions that does not fit neatly with the notion of separate "emotion" circuits—we can't isolate particular emotions to one region of the brain. We know from modern brain-scanning techniques that many different areas of the brain become activated at the same time during emotional *and* thinking experiences. If I could have peered into my brain as I was sprinting away from that dog, I would have seen not just my antiquated amygdala going ten to the dozen but also many other areas in a heightened state of activation. Also, as our brain evolved it did not proceed in a logical linear way, as the triune brain idea suggests. Instead, just like an organization, such as a company or a university, restructures as it expands in size and complexity, so the brain constantly reorganizes and adapts as it evolves.

What this tells us is that the dense interconnections between the so-called "thinking/problem-solving" parts and the "emotion" parts of our brain allow them to work together in a seamless way. In fact, assemblies of cells across the brain can connect rapidly in a kind of "emergency response callout formation" to help us deal with specific situations, and these connections have no respect for borders. Instead, we now know that the brain works as a highly integrated and dynamic system that operates as one.

Of course, within such an integrated brain it's still possible that separate assemblies of brain cells—usually called "circuits"—for different emotions might still exist. It would make sense, wouldn't it? In my encounter with the aggressive dog while running, unique *fear* circuits might have come into play to help me escape. However, this too may be wrong. Surprising though it may seem, it is proving difficult to find convincing evidence for distinct brain circuits underlying fear, anger, or disgust, or any other emotion for that matter.

Describing our emotions is not as straightforward as we might think

Not only are emotional circuits difficult to pinpoint on brain scans, but when we ask people to describe their emotions these descriptions also do not fall neatly into the familiar categories of "fear," "sadness," "joy," and "disgust" that we might expect.

Think about it. Try to put into words what it feels like to be afraid. Now explain what it feels like to be angry. If you take away the *cause* of your feeling, can you really describe how fear feels different from anger? Difficult, isn't it?

Many studies in psychology have shown that it is, in fact, virtually impossible. Instead, what people do describe are much broader *dimensions* of affective experience along the lines of how intense a feeling is and how negative or positive it feels. Results like these began to alert researchers to the notion that the classical view of emotions may not be correct. If I try to describe my fear experience with the dog, or even try to describe the feelings of terror I had when clinging on to the diving pillar all those years ago when I was in fear of drowning, it is difficult to get away from physical descriptions. On both occasions, I remember my heart hammering in my chest, I felt wobbly, my mouth went dry, and afterward, in both cases, I experienced an uncontrollable trembling. In one experience I became rigid and stuck, in the other I ran as fast as I could. But in terms of what it *felt* like? I can say it didn't feel nice and that it was intense, but it's hard to describe beyond that.

I'm not alone in this. This is what is typically found in emotion research.

The building blocks of our emotional life: arousal and feeling tones

The foundations upon which we build our emotional life seem to be based on the broad dimensions of how *arousing* a situation is and whether it is *positive* or not, rather than the set of discrete emotions that seem so intuitive. Judging whether a situation or object is negative or positive—called a "valence judgment" in psychology—is close to the idea of "feeling tones" that we mentioned back at the beginning of the book. Mindfulness teachers tell us that a sense of whether something is pleasant, unpleasant, or neutral—"feeling tones"—give us an indication of what's important. This is consistent with contemporary emotion research that shows us that feelings are a kind of tracking system in the brain alerting us to what feels good and what feels bad, so we know which experiences to avoid and which to approach.

The important point is that brain circuits involving *both* cortical and subcortical parts of the brain maintain and strengthen negative or positive interpretations of everyday events. The results confirm that the brain operates as a highly fluid connected system drawing on many different regions to deal with emotional situations. We have seen in Chapter 6 that cognitive flexibility when dealing with emotional situations is especially important in keeping our mind running fluidly. The new emerging view of the brain as working in a highly integrated way provides us with a deeper understanding of why emotions and our awareness of our emotional life are so important for agility, and switch craft more generally.

Emotional switch craft

The emerging view of the brain as a prediction machine—and one that works as a highly dynamic unit—tells us that we do not need several inflexible brain circuits for a functional emotional life. Instead, a minimal number of general processes are sufficient. A process that computes the valence of an external event (is it "good" or "bad"), an ability to categorize that event rapidly, and an ability to integrate these processes with internal bodily information is all that is needed to help us react appropriately in the moment. This is the essence of the idea that our emotions are constructed "on the hoof" rather than being built-in.

What this tells us is that changes in the body are transformed into an emotion when they acquire psychological functions that they cannot perform on their own. Our emotions, in other words, emerge from a combination of three things: a highly flexible brain, an intimate understanding of the environment within which it operates, and the meaning of internal signals coming from the body. This view is drawn from a much broader understanding of how the mind works. All mental states, according to this standpoint, are created in those moments when our thoughts, our feelings, and our perceptions all come together in a single moment. The important point is that this temporary fusion is tailor-made for a specific situation and is created by drawing upon prior similar experiences that help to shape our response to this new situation. This theory of constructed emotion offers us a very different way to think about emotions.

Take my experience of fear when running away from that dog. By this theory, when I saw the dog charging toward me, basic survival circuits in the brain indicated that immediate action was needed (*arousal*) and that the situation was bad (*valence*). At the same time my thinking brain categorized the episode as "potentially dangerous and perhaps fearful." Using elements of my past experience, when I had been chased previously by an aggressive animal—a crab chasing me along the water's edge as a child is a particularly memorable example—my brain speedily predicted what my body needed to do to cope with the situation. It was those *predictions* that caused the flood of adrenaline, which allowed me to sprint away and escape the danger. My subsequent labeling of the experience as "fearful" then helped me to give some meaning to the various sensations I was feeling as an instance of *fear*.

These observations force us to reconsider all that seems so intuitive about our emotions. In personal terms, as a scientist who has trained and worked for many years within the classical tradition, it has taken me a long time to come to terms with this new view. While it often just doesn't feel right, the growing evidence makes it harder and harder to disagree with the idea that emotions are largely constructed rather than being prepackaged for us by nature.

The view that emotions are constructed, rather than biologically given, suggests that they are produced through a process of categorizing physical changes in the body—such as increased heart rate—in relation to the current situation. Your heart rate may increase when you are sexually excited, when you do a high-intensity training session, or when you are sprinting away from an aggressive dog. In each case the physical change (increase in number of heartbeats per minute) is often identical but the way you interpret the situation is entirely different. This gives us a very different perspective on emotions. Rather than being hardwired, they are created flexibly to help us cope with the demands of rapidly changing events. This shows us very clearly why emotions are essential for agility. If emotions are constructed in this way, rather than being hardwired as the classical view suggests, then this provides us with a window of opportunity to manipulate and change our emotional responses in the service of switch craft. As just one example, this allows us to modify our feelings by modifying how a situation is interpreted. Reframing an upcoming public talk or presentation as an interesting challenge, rather than as a threat, for instance, can actually change the emotions that are constructed.

Nature often creates broad solutions that can help to solve many different problems. The *constructed emotion* view, like the classical *biologically basic emotions* view, is based on evolutionary assumptions. However, what has evolved is different. Rather than numerous specific circuits for each emotion, a smaller number of more general processes are important. The idea is a straightforward one. Rather than developing a specific solution for every specific problem, which would be highly inefficient, nature often develops a small set of processes that can then be used to solve a wide variety of quite different types of problem. Such general processes are much more efficient and allow for greater flexibility than do processes that are specific to particular situations. Again, the constructed emotion view suggests that understanding how emotions are made is critical to understanding how we can remain agile.

There is a long tradition of this way of thinking in psychology. The assumption is that how we perceive objects, what our attention is drawn to, how we memorize and categorize things, and even how we learn new things are broad processes that apply across many settings. A good example is the well-known limitation on our ability to remember things in the short term. We know that people are typically able to remember around seven items, plus or minus two. This means that if we give people a list of twenty items and ask them to recall as many as they can, the average number recalled will be seven, with most people remembering between five and nine items—seven, plus or minus two. Whether the list consists of Pokémon characters, items in a supermarket cart, words, or numbers, the limitation is the same—there is a general constraint on short-term memory regardless of content.

What has emotion ever done for me?

There is, of course, a fundamental question that arises at this point of the discussion. Does knowing where emotions come from, what factory—"hardwired" or "constructed on the fly"—they are made in, provide us with any insight into what they are actually for? Emotions, after all, are the most subjective element of our consciousness.

Although we might describe similar emotional experiences—the special weightlessness that comes with joy, say, the clenching of the stomach that comes with apprehension and dread, or the relentless shakiness that comes with anxiety—no one else can truly understand how you feel. As we've seen, scientists don't all agree on how emotions are made—some believe there are inherited emotion circuits in the brain, others that emotions are transient combinations of different processes that are put together in the heat of the moment. But the strong feelings we experience when we are in love, terrified, or in deep grief can be more real to us than any such explanation.

To answer the "why" question of emotion, let's step out of the laboratory for a while and immerse ourselves in everyday life. Firstly, one thing we do know is that emotions provide us with internal information from our bodies. Our feelings act as signals to us that all is well or that all is not well—what the mindfulness tradition calls "feeling tones." They release powerful thoughts and images that infuse our very being. These can be disruptive and stop us in our tracks or they can be encouraging and motivate us to continue. Bodily feelings conspire with powerful thoughts to elicit action, encouraging us to behave in the most appropriate way for the situation. Imagine being held up at knifepoint by a mugger as you walk down a dark deserted street at midnight. For most, the overriding sense of fear would "persuade" us to hand over our valuables and get out of the situation as quickly as possible. But for a Special Forces soldier friend of my husband, precisely such an event constituted a "call to arms," a "fun" interlude on his way back home from the pub. In no time at all he had disarmed the mugger, pinned him to the ground in an armlock, and called the police on his mobile phone to get him picked up. His overriding sense hadn't been fear, it had been excitement—and the result had been very different.

Different emotions can inspire a variety of actions, again giving us a lot of agility in how we can react in certain situations. Fear can induce you to flee, fight, or perhaps freeze. Sadness can motivate you to withdraw, recover, or begin the process of moving away from a cherished goal. Joy might encourage you to approach and maintain the pleasant state you find yourself in, while disgust provokes you to avoid something rotten. In reality, of course, our feelings cannot be so neatly categorized, and we can feel many competing things at any one time, but these emotions are broadly experienced as feelings of either pleasantness or unpleasantness—the "feeling tone"—and they help us decide whether to persist or change tack.

Negative emotions are useful

All of your emotions, even the very unpleasant ones, are important to achieving psychological health and happiness. Negative emotions, like anger or fear, are associated with threats, and narrow your attention to help you to focus on important issues that may harm or hinder yourself or your loved ones. This is one reason why they can feel so overwhelming at times.

Unsurprisingly, perhaps, negative feelings tend to shout louder for your attention and can be "high maintenance." Feeling bad encourages you to avoid situations that produce such feelings, which can be a good thing. If a particular group of friends always leaves you feeling low and angry, for instance, then you might sensibly reflect on this and decide to spend less time with them. That said, however, you shouldn't avoid the unpleasant feelings themselves. If you persistently avoid stress or discomfort, then you will limit your horizons and you may not be able to achieve what you want.

When we look at people who have been resilient through difficult times and who have achieved what they wanted in spite of setbacks, one thing they have in common is an ability to cope with feeling bad as long as this is in line with longer-term goals. If the thought of asking for a promotion is terrifying, for instance, you could make yourself feel better by deciding not to ask. This might be a relief in the short term but won't do your career prospects much good. Just as an athlete won't enjoy getting up to train on a cold winter's morning, they know it's important to get used to it to achieve their goals.

To return to our gearbox analogy, imagine that I were to erase all fear from your brain for thirty minutes and then sit you behind the wheel of a car. How long do you think you would last? What would stop you from overtaking the old lady driving at a snail's pace on a blind bend? Next time you arrive home safely after a drive, remember that fear had a lot to do with it. Similarly, imagine if I were to erase all sense of anger and frustration from your brain and then sit you down at a negotiating table. How much of what you want do you think you'd end up leaving with? Negative emotions like fear, anger, and disgust are important because they force us to focus on events and things that pose a threat. This is why what I have previously called the "rainy brain"—those brain processes that alert us to danger and threat—has a much stronger pull on our attention than more positive and rewarding events.

The benefits of positive emotions

While negative emotions are essential, life is of course better when you also regularly experience positive feelings. I am talking about a range of distinctive positive emotions rather than just a general feeling of happiness (some of the most common are joy, gratitude, serenity, interest, hope, pride, amusement, inspiration, awe, love, and curiosity). Positive emotions tend to widen our attention, broaden our thoughts, and urge us to be inventive.

Positive emotions are energizing because they encourage us to want more of the same. The mechanism by which this occurs in the brain is somewhat surprising. Positive experiences trigger the reward center in our brain—the *nucleus accumbens*—in different ways. This reward center can be divided into two parts—one that makes us *like* things and another that makes us *want* them. The "liking" parts release hormones such as endorphins that are naturally occurring opiates and provide us with feelings of pleasure ("liking"), whereas the "wanting" parts release the chemical dopamine, which makes us pursue more of the same ("wanting") and bolsters our ability to persist.

Wanting, crucially, does not always go hand in hand with liking (which is why we don't always like what we want). Many professional athletes I know want to train but don't particularly relish the prospect. Most drug addicts get to the point where they want their fix but hate it. Indeed, a forensic psychiatrist friend of mine told me that many pedophiles detest giving in to their urges.

Positive emotions are powerful motivators, then. Many studies have shown that positive experiences open us a little, they widen our attention and enhance our sense of wonder. When you feel positive your interests become broader and your creativity increases. And they not only broaden our attention but have also been shown to enhance our ability to switch tasks.

When you are in a good mood, your thought processes are also likely to become more thorough, which can lead to marked improvements in your decision-making. In one study, some medical students were either given an unexpected gift or played uplifting music, which led to an increase in positive mood. The students in a control group received no gift or were played neutral or slightly downbeat music that had little impact on their mood. The students who were in a good mood came to a correct decision more quickly, and showed less confusion in coming to their decision when asked to make a diagnosis based on several key symptoms. And even experienced doctors were found to be more efficient in diagnosing specific illness and were less likely to be "anchored" to their original expectations when they were in a good mood.

Put simply, the upbeat doctors were more open to new information, even information that conflicted with what they were currently thinking. Their attention was widened, and their mental agility increased, as they flipped back and forth between different aspects of the situation. And this isn't just the case in the medical profession. Positive emotional experiences can guide all of us into taking multiple aspects of a situation into account so that our evaluations are more responsive to the circumstances and less susceptible to bias.

Positive emotions support resilience

Positive emotions and resilience are also closely linked. After the terrorist attacks of 9/11 in New York City, people unsurprisingly experienced a variety of emotions: feelings of anger, fear, and hostility toward the terrorists were common. However, some people were able to snatch moments of joy and connection with family and friends in the midst of despair, as well as feelings of hope and inspiration for the future. Those who could do this even fleetingly showed more resilience in the months that followed. So, when you are dealing with a crisis remember to try to seek even tiny positive experiences; it might be playing with your baby, phoning a friend, or indulging in some of your favorite chocolate.

Positive experiences and emotions can also be "banked" to draw upon when times get tough. If you regularly experience positive emotions, your social bonds will strengthen, and your resilience will increase naturally as you learn to face difficult situations with a broader perspective and a willingness to be agile. Years of painstaking research have shown that if you experience higher ratios of positive to negative emotions (3:1 is the often quoted although much disputed minimum ratio), you will be better able to navigate everyday difficulties and challenges.

How do we boost our positivity?

The science of positive emotions shows us many ways in which we can become more positive, even when we are dealing with a lot of difficult stuff in our lives. The important thing to remember is that small yet frequent doses of positivity can work wonders. However, what is clear is that *trying* to be positive when you don't feel that way can backfire and lead you to feel worse and miss out on the benefits of positivity. We all know people who act in a falsely upbeat way and are unrealistically positive. Trying to generate feelings of positivity when they are just not real is not worth the effort. Instead, here are some mottos that will be more helpful in generating positivity without having to work too hard at it:

- **Be grateful:** This is one of the easiest of the positive emotions to generate. Ask yourself what is going on in your day that is a gift, what are you grateful for? It may be something as simple as a sunny day, it may be your dog, or a great group of friends. Finding something that you are thankful for is an easy way to give yourself a genuine dose of positivity.
- **Be curious and open:** Being open and curious is a great way to help you shift away from all the negativity and focusing on what's wrong to seeing possible benefits in a situation. So try to be curious, even when you don't feel like it.
- **Be kind:** It is often not difficult to show kindness to other people and to animals, and lots of evidence shows that simple acts of kindness can give us a warm feeling of well-being that helps our overall positivity.

- **Be appreciative:** If someone helps you, or is kind to you, let them know. Show them that you appreciate what they have done. This will not only give them a boost of positivity, but will also make you feel better.
- **Be real:** Don't try to fake positivity. Acknowledge to yourself and others if you are feeling really bad and going through difficult times. Just be careful that you don't wallow too much in the negativity and keep it in. If you have had a major setback, wallowing for a time is OK, but after a while try to find some things to keep you occupied that will give you a focus to help you move out of the bad mood. This is much more effective than trying to express the negative feelings.

Positive emotion, then, is more than simply feeling good. Positivity opens us to new experiences and relationships and sparks our curiosity and creativity. Like a flower that opens toward the sun, we become exposed to the amazing diversity of life. From this place of openness, you will be able to build lasting resources—things like friends, a variety of experiences to draw from, or a sense of purpose and meaning in life—that will remain with you long after the emotion has passed.

And *that* is "good" news for all of us!

Chapter Summary

- Emotions are an important pillar of switch craft because they allow us great flexibility in responding to challenging or changing situations.
- All of our brain regions are highly interconnected.
- Our emotional experiences are probably constructed rather than hardwired.
- All of our emotions, negative and positive, are important because they provide us with the vital "feeling tones" that tell us whether all is well or not.

- Emotions are the gears linking thoughts and judgments to actions.
- While negative emotions aid our survival, positive emotions should not be underestimated. They are vital ingredients that facilitate thriving and boost our resilience.

CHAPTER II

LEARNING TO REGULATE YOUR EMOTIONS

I once heard a story that amused me about a famous professor who gave a keynote lecture on emotion regulation at a major international conference. At the end of the lecture, during questions, a rather portly member of the audience stands up. "Could you tell me precisely *why* emotion regulation is so important?" he asks the professor. The speaker stares at him in stony silence. "No, I couldn't," he replies. "Now sit down, you fat bastard!" There are gasps from the audience. Did they really just hear him say that? The inquisitor himself is incandescent. "How dare you talk to me like that in front of all these people!" he rages. "I spend an hour of my time here listening to you ramble on, ask you a question in good faith, and all you can do is insult me!" Instantly, the speaker composes himself. "Please, sir," he says, "forgive me. I have absolutely no idea at all what came over me just then. I am mortified. To make up for it, I will give you a free copy of my book at the end of the talk and buy you a whiskey. I think I need one myself! I am so sorry. I beg you to accept my apology?" Sufficiently placated, the man accepts his apologies and settles back down in his seat. The speaker pauses for a moment before continuing. "Now," he says with a smile, "does *that* answer your question?"

Keeping our emotions in check

While it's good to tune in to our emotions and be open to what they are telling us, sometimes the intensity of our feelings can be overwhelming, and we need to find ways to bring these strong emotions under control. Day-to-day life is filled with attempts to control our emotions. This is really important for switch craft. In order to calmly assess situations and decide on the best approach to tackle whatever problem we are dealing with, we need to be able to see and think clearly, without being blinded by strong emotions pushing us to act on impulse. Emotions are, of course, often good for us and help us to adjust and shift between situations—vital for agility. But they can also be overwhelming, so to be truly agile we need ways to keep them in check when they are hindering rather than helping the situation. My fear was not helping me when it was encouraging me to cling to that diving pillar all those years ago; instead, I had to regulate my fear in order to brave the waves and swim to safety. A person in the grip of passionate love may not always make the best judgments about the person at the center of their desire.

We can learn a lot about how to regulate strong emotions from one of the so-called new generations of talking therapies called "dialectical behavioral therapy," or DBT for short. This form of therapy combines an emphasis on helping you to change unhelpful ways of thinking with a focus on accepting yourself as you are. "Dialectical" means trying to understand how two things that seem to be contradictory—such as accepting who you are, but also wanting to change your behavior—can both be true at the same time.

Take an example of a soccer player I helped deal with heavy alcohol use. He confessed that he felt the pressure of competition and was nervous about admitting this to his coach, so he had turned to drinking to help curb his strong anxieties. While he had hidden it well, his drinking was beginning to escalate, and he had also started to use sedative drugs at the same time. Working through the pressures he felt, we both came to the realization that his behavior made absolute sense in terms of trying to reduce his anxiety and stress. This was the acceptance part—rather than judging his behavior, we agreed it was an effective way he had found to cope with his stress, at least in the short term. At the same time, we both agreed that this behavior was not helpful in the longer term and would soon begin to affect his performance. It was clear that he needed to find new ways of dealing with stress.

To do this, we explored things that he really enjoyed and discovered that he had always found cooking very relaxing. Buying fresh ingredients in a local market and cooking had always relaxed him, but this was something he'd stopped doing when he moved cities to join a new club. He arranged to have fresh ingredients delivered and began to reignite his love of cooking. This also allowed him to become creative and invite some friends around for meals. While he did not stop drinking totally, he found that he became less dependent on drinking to relax and began to replace this activity with preparing, cooking, and eating good food. A positive side effect was that he also found he began to sleep much better, and this also helped him to cope with feelings of stress.

A tip from therapy to help us control intense emotions

Dialectical behavioral therapy (DBT) suggests the following method to deal with any difficult situation—work out which "mind" you are in. The *emotional mind* is where you read the situation through your emotions and feelings. The *rational mind* is where you understand a situation through the facts and the figures. Then there is the *wise mind*, which meshes the emotional and rational minds together. The key is that the wise mind asks, "What do I need to do in this moment that's aligned with my values?" or, to put it another way, "What is my truth in this?" Simply asking yourself "What mind am I in?"

allows you to gain some degree of control over the situation. Then ask yourself: "If I was in wise mind, what would I do?" The wise mind is of course the state of mind most likely to allow your switch craft to operate effectively. So nurturing a wise or a "switch craft" mind is hugely helpful in dealing with difficult situations.

DBT has also developed a useful mnemonic to help people cope better with future unpleasant emotions and to build emotional resilience. We can all use the ABC-PLEASE technique to help us cope with and bounce back from stressful experiences.

- Accumulate as many positive experiences as you can
- Build your competence by learning skills that you enjoy
- Cope well by doing your background research and making a plan
- Physical illness—if you are sick or injured get proper treatment for it
- Lower your vulnerability to poor health as much as possible
- Eat healthily—make sure you eat enough and feel satisfied
- Avoid (nonprescription) mood-altering drugs
- Sleep—ensure that you do not sleep either too much or too little
- Exercise regularly

These are very general life rules that will bolster your switch craft by ensuring you are alert, have plenty of energy, and are capable of regulating strong emotions. Being physically and mentally prepared is essential for switch craft so these very general rules of living are vital.

Sometimes more specific strategies are needed

There are times, of course, when you need more specific strategies to help you regulate your emotions. Thankfully, there are many actions you can take to regulate specific emotions. When making an important presentation at work, you will probably want to reduce your feelings of anxiety. If you have recently lost a loved one, you may want to lift some of the intense sadness you feel before meeting friends. Although our emotions can feel uncontrollable, there are many ways to influence your emotional state. Rethinking a difficult situation to reduce stress, breaking it down into more manageable components, or listening to upbeat music to lift a low mood are all examples of emotion regulation. Finding ways to change how you think and feel about your changing circumstances can make the difference between responding well in a crisis or succumbing to panic and anxiety. Developing good emotion regulation skills is vital, especially when dealing with ongoing stressful situations, as the coronavirus pandemic was for so many of us.

How to regulate your emotions

We still know surprisingly little about how people choose to regulate their emotions and this is a burgeoning area of research. To learn how to regulate your emotions you need to make a few decisions.

- **First, ask yourself if emotion regulation is required.** Do you need to dampen down feelings of anxiety? Lift a low mood? Reduce your degree of excitement?
- **Second, choose which regulation strategy is best.** You might ask yourself whether it's possible to change your situation, or whether you can distract yourself. If you are at the dentist, avoidance is usually not a sensible option, so perhaps diverting your attention by listening to some favorite music might help. Research has identified four general types of strategies that we can use: changing your situation, changing the focus of your attention, changing

how you think about the situation, and changing how you respond.

- **Decide how to put your chosen strategy into action.** Keep a close eye on things to decide whether to stick with your chosen strategy, switch to another approach, or stop the attempts to regulate altogether.

Emotion regulation is an ongoing process and a central part of switch craft. The diagram below illustrates the four general types of strategies that can be used.

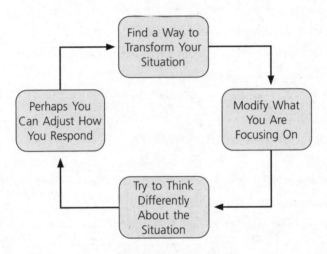

Sometimes, changing the situation, if possible, is the best approach. Hence, I ran as fast as I could from that aggressive dog. At other times, you cannot avoid the situation, and so equally important is strengthening certain mental processes, such as how you interpret situations or modify the focus of attention, that help to regulate emotion in future events. The ability to find different ways to interpret and reframe stressful events can be especially helpful. Finally, there are several useful things you can do that can directly affect your emotional response. These might include things like slowing down your breathing or taking a stimulant to lift your energy levels.

The table below outlines some specific strategies that are commonly used to help us deal with stressful situations and strong emotions. Most are helpful, while some—such as rumination and worry—can all too easily get out of control.

Find a Way to Transform Your Situation
You might be able to avoid a situation entirely.
Perhaps, you can use humor to lighten the situation.
Seek out friends or colleagues who can support you and therefore make the situation easier.
Sometimes, simply turning off your mobile phone can help.
Modify What You Are Focusing On
Start paying attention to your breathing—count the number of seconds as you breathe in and the number of seconds as you breathe out. Doing this for a few minutes can help to "ground you."
Thinking about something else can help to distract you from an emotional situation.
Rumination or worry can distract you, although they also often make the situation worse. But there are times when thinking repetitively about why you are feeling the way you are (ruminating) or figuring out what might go wrong in the future (worrying) can be helpful.
Try to Think Differently About the Situation
Reinterpreting the meaning of a situation by stepping back and seeing the bigger picture, or perhaps looking for the silver lining, can often be helpful. This is formally called "reappraisal" in psychology jargon.
Another useful technique is called "cognitive distancing" in which you take a third-person perspective to look at yourself and how you are dealing with an emotional event.
Accepting a distressing situation and allowing yourself to experience the negative feelings that go with it can be very helpful.

Perhaps You Can Adjust How You Respond

Drinking some alcohol to take the edge off anxiety or to help you relax, or perhaps drinking some caffeine or an energy drink to give you a temporary boost can help in the short term.

Inhibiting how you are feeling, such as smiling even when you don't feel like it, can sometimes be effective.

Once again, deep breathing can be useful. For example, taking a few deep breaths can help to calm you down if you are feeling very angry.

You might try to get more sleep if you find you are overreacting to situations because you are overly tired.

Exercise often helps. You might go out for a run or head to the gym to give you more energy or calm you down.

To bring this to life a little bit, let me tell you about a successful businesswoman I once coached called Mandy. Mandy sought my help for her chronic worrying—though her problem was less about actual anxiety and more about emotion regulation. To all intents and purposes, she had an enviable life. She had a good marriage, two great children who were doing well in school, a wide circle of friends, and she loved her job in a large architectural firm. Her role was to obtain lucrative new projects for the company. It was demanding but also highly rewarding. By all accounts, she was very good, regularly winning large contracts for high-profile building projects. I sometimes saw her name mentioned in the newspaper associated with some exciting new property development in London.

Mandy didn't have a 100 percent record—who does?—and she couldn't help thinking about the contracts she didn't win. Over and over, she would ruminate on whether she had done something wrong. Had she not represented her company well enough? Why had she failed to convince her potential clients that she could provide a better all-round package than competing firms? As soon as Mandy started thinking about her work, her mind was caught in a vortex of negative thoughts of all the times she had failed, and her negative emotions were overwhelming. Rather than cherishing her more frequent successes, Mandy was laid low by her occasional failures.

We worked together on several strategies from each of four groups to help. Mandy realized that she was using worry as a way of keeping her mind on the things that had not worked out—this focus on the negative had to be altered (*modify what you are focusing on*). So, she changed the situation to some extent by sharing her worries with some of her colleagues and often recounting funny stories about why things had not worked out (*transform your situation*). She also actively tried to reinterpret the meaning of the situation (*think differently about the situation*) by looking at the bigger picture and realizing that nobody had a 100 percent record. Rather than focusing on the failures, she learned to celebrate her successes and put them in context. Finally, she also made sure that she got regular exercise and better sleep so that she did not feel so stressed, and she also learned some deep-breathing exercises that she could do any time she began to feel stress (*adjust how you respond*). Over time, Mandy learned to manage and regulate her worries and her negative emotions.

As you can see, emotion regulation is a continuous and ongoing process, just like switch craft itself. We don't have an emotion and then regulate it. Instead, we experience multiple emotions continuously, which means we are constantly not only listening to our emotions, but also making decisions about whether they need to be regulated and, if so, choosing the best approach. As in Mandy's case, a variety of different approaches is usually the best solution.

Many of us can be disrupted by negative emotions and have difficulty in regulating our negative thoughts, which can undermine our switch craft skills. We have a natural tendency to laser in on negative information. On the one hand this is understandable. Our brain will always magnify potential danger more than potential reward, because it was once far more important for our ancestors to notice a threat above all else in order to survive. Negativity is catnip for most of us. The problem arises when these thoughts become a habitual response or turn into what clinical psychologists call "Automatic Negative Thoughts" (ANTs), when our knee-jerk reaction to every situation is a negative thought, which spins into another one and then another—when, in other words, we fail to *regulate* that negativity. These ANTs are personal, pervasive, all-consuming. As my client Mandy discovered, they can infest everything in your life and undermine your happiness.

The important thing to remember is that these negative biases are not always bad. Sometimes being alert to potential danger is essential. This is why all of us will have negative thoughts and these do not inevitably lead to depression and anxiety. It's not the bias itself that is the problem but the *rigidity* with which people apply it. One of the key features of a mind trapped in the grip of depression or anxiety is the tendency to become stuck in these habitual and repetitive ways of thinking. The constant chatter in our head does a great job in keeping us stuck. So, an important aspect of emotion regulation is often finding ways to stop your negative thoughts from getting out of control. Nothing torpedoes your switch craft as much as a mind caught in the never-ending loop of worry and rumination.

How to manage negative thoughts

It's important therefore that we learn to manage negative thoughts, working out when they are useful and when they aren't. There are many ways to reframe how you look at things. Let's say you want to become more assertive at work. Think carefully about why you are not more assertive. Is it because you're worried that you will be seen as "pushy"? Think about how you might rethink these unhelpful thoughts. Perhaps write down a couple of your beliefs—"People won't like me if I become more assertive" or "My boss will think I'm arrogant if I ask for a raise," for example—and then interrogate them by asking the following questions:

- Is the belief black-and-white, and without nuance?
- Is the belief all-encompassing? Words like "always" or "never" are danger words.
- Does the belief assume that you know what others are thinking?
- Does the belief focus on the negative?
- Does the belief contain an element of what you feel you "should do" or "ought to do"?
- Does the belief blame other people? Or are you cast as the victim?

It is surprising how a belief can lose its power when you work your way through these questions. Turning detective and interrogating your beliefs like this often reveals that they are based on your own set of heavily biased assumptions rather than on reality. Reframing "I am too pushy" as "I am trying to achieve my goal of getting a promotion" or "My request for a raise is reasonable" lays the groundwork that will help you to start challenging your perspective on a regular basis, and this will ultimately lead you to become less rigid and more agile. You need to zoom out and resist the tendency to become overly focused on your problems. Rather than having your nose to the painting, as it were, and dwelling on every tiny detail of *why* this has happened, zoom out and start to think about *how* you can resolve the situation.

Change your "why" questions to "how" questions

Many studies have told us loud and clear that there is nothing more guaranteed to make you unhappy than ruminating on why something bad has happened to you. This negative chatter in your head—Why did I get cancer? Why did my boyfriend leave me? Why did I not get that job?—can be an impossible loop to escape from. So, any time you spot these "why" or "if only" thoughts, ask yourself some "how" or "what" questions: How can I make myself feel better? What can I do right now to change the situation?

Clinical psychologists find that this simple technique works wonders. People who are struggling with post-traumatic stress disorder are often consumed by thoughts of why their accident or trauma happened, and this only succeeds in keeping them stuck in their negative thinking. Spotting this and replacing it with a "What can I do to move on with my life?" question can magically transform people by depriving this cycle of negative thinking of its essential fuel. Once you start to focus on the "how" and "what can I do" questions, the "why" questions seem to lose their power and you can begin to get out of your head and back into your life.

Reframing a situation is also a powerful technique

Reframing, sometimes called "cognitive restructuring," can help you to control how you react to different situations. Let's see how this works by looking at another real-life example. John was a fit and healthy thirty-eight-year-old man who had been working as a security guard at a large factory on the outskirts of town. He usually worked the night shifts; on most nights, not much happened, so the most difficult problem he had to face was boredom and trying to ensure he didn't fall asleep. One night, however, three men attacked John at work. One of the men punched him, while another pointed a gun at him and told him to lie on the floor. Terrified, John had stayed put until he was sure that the robbers had left. Two years after the robbery he was still suffering with severe anxiety, often too scared to leave his house, and too scared to work nights.

I asked John to keep a diary of his negative thoughts and beliefs for a week. When we looked over this diary together a week later, it became apparent that most of John's negative thoughts related to a fear of being assaulted again. I asked him to estimate the chances of him being attacked again at work and he rated this as about 80 percent. I then encouraged him to challenge this belief using the laws of basic probability.

"How many times have you worked the night shift?" I asked.

"Hundreds of times, at least," he replied.

"OK, so let's say two hundred times," I suggested. "Now, on how many of those previous occasions have you been attacked?"

"None," he replied.

"OK, what about the other security guards at the factory—how many times have they been attacked?"

"Once in the last ten years," he said.

As we continued to explore this, it became obvious to John that he was vastly overestimating the likelihood that he would be attacked again. It *could* happen, of course, but it was clearly a very rare event. When I asked him to re-rate his chances of being attacked, he now suggested it was probably around 1 percent, a far cry from the 80 percent he had slipped into believing previously. John slowly began to reframe his negative belief by acknowledging that his chances of being attacked were actually very low and no more than other people's. This new belief, this reframing, reduced his anxiety and helped him to start living a normal life again. He could now think about the *how* question and work out ways to start enjoying his life.

Being flexible about what emotion regulation technique to use is key

Reframing our feelings is often contrasted with the *suppression* of our feelings, something many of us try to do automatically. This might be telling a group of children who are out in a small boat in a storm that everything is going to be fine, even though you are feeling very nervous yourself. We often think of suppressing feelings as being unhealthy, and studies have shown links with psychological difficulties. However, as with many things in psychology, and life, it's not quite as simple as that. One of the most important lessons we can learn from switch craft is that the effectiveness of any strategy very much depends on the situation.

Indeed, some studies have found that it is the flexibility with which we use different emotion regulation that really counts, not any particular strategy itself. For instance, one study was conducted that followed a group of one hundred eighteen-year-old students for two years following the 9/11 terrorist attacks in New York City. Soon after the attack, the students came into the laboratory and were asked to look at a series of highly positive or negative emotional images. For some of the images they were asked to "express the emotions you feel as fully as possible," while for others they were asked to "suppress as fully as possible any emotion you feel while viewing the images." Each student was filmed while doing this and told that another person would have to guess if they were feeling an emotion or not. So, for both the expression and the concealment conditions they had to try to ensure that someone viewing their video would be able to work out whether they were feeling calm or anxious. Some were really good at suppressing what they were feeling, others were great at expressing their emotions, while some were really good at both and were able to easily adapt according to the instructions—to either conceal or express. It turned out that flexibility in the use of *both* strategies was most important for resilient functioning. Remarkably, those students who were better able to agilely express *and* conceal their emotions in this simple task were less distressed when they were tested again up to two years after the terrorist attacks, compared to those who favored one or the other strategy.

The important lesson here is to apply emotion regulation strategies flexibly in a way that fits best with the nature of the challenge you are facing. Choosing the right strategy is important and our experience is often the best guide. That's why your body provides moment-to-moment updates about the world around you and why these signals prompt your brain to make continuous adjustments to your behavior. So it's important to learn to listen to your body and your emotions; these are key skills required by any switch craft practitioner. If you feel cold you might have a cup of warm soup; if you feel tired you might stop what you're doing and get some rest. Our emotional regulation needs to be constantly adaptive in the same way. We need to be *agile* on the one hand, to adapt our emotions according to the situation; and we need to be *authentic* on the other: we have to be true to ourselves.

Accept ourselves as we are—an effective way to manage our emotions

One powerful therapeutic approach built on both principles of agility and authenticity is called "Acceptance and Commitment Therapy" (ACT). ACT encourages us to work with ourselves as we are, rather than desperately trying to change ourselves. The key principle is simple: your actions must always be guided by your core values—those things that provide meaning in your life. ACT is all about taking *values-based actions* to inspire real behavioral change. The idea is to:

- Accept your thoughts and feelings and be present in the moment
- Choose a way forward that is consistent with your values
- Take appropriate action

This is another reason it's important to establish your values and goals (as we have spent some time working on earlier in the book). The idea is to focus on accepting negative feelings and thoughts as a normal part of your life, and putting your energies to better use in engaging in activities that you really value.

Your thoughts are not descriptions of reality

To "get out of your head," regulate your emotions, and become more agile you must learn to see that your thoughts, no matter how powerful, are not actual descriptions of "reality"; instead, they are symbols of your own personal experience. That's a big difference. Your thoughts are important to understand who *you* are, but they don't necessarily give you a factual understanding of the outside world. People have all sorts of weird and wonderful beliefs that bear little relation to what's really going on. Falling into the trap of believing that your negative thoughts are telling you the unadulterated truth can lead to untold misery.

Alan, a man I was coaching, explained to me that he had experienced low moods and lack of motivation for many years. "It's a physical problem," he assured me, "not mental." He had been in a car accident when he was five years old and, although he was uninjured, Alan was convinced that an unidentified brain injury occurred at that time that explained why he wasn't motivated and didn't enjoy life. His solution was to spend most of every day piecing his life together, using rumination and worry to try to figure out when he'd first noticed his problems. He was convinced that the accident was the key to why his brain "worked" differently from everybody else's.

Because of this belief, Alan's life had been on hold for many years. He'd had brain scans, spoken to neurologists, psychiatrists, and psychologists. None had found anything wrong. I gently suggested that the main problem with his brain was not physical, but rather the repetitive negative thinking itself. I showed him several studies demonstrating that rigid and fixed beliefs like his, along with constant rumination about them, are common and unsuccessful ways of coping. While he agreed with the research, he insisted it was different for him; he was ruminating, he believed, *because* he had a physical abnormality with his brain. "Working on the rumination itself," he said, "won't help."

In the end, what did help Alan was to take a more agile approach and learn to accept that his belief about his brain being damaged was simply a thought that may or may not be true. Once he began to genuinely consider the possibility that his belief might be wrong, he was on the road to recovery. Of course, taking such a step is easier said than done. In Alan's case, the breakthrough came only after two years of coaching alongside working with a cognitive-behavioral therapist.

Sometimes, like Alan, we need to accept that ruminating as a strategy to work out why you are struggling in life is not the most effective approach. Sometimes, as in his case, the breakthrough comes when you accept that whether your belief is right or wrong is not necessarily what's important. Rather than trying to figure out exactly what might be wrong with your brain, say, a better approach is to look for ways to help you live a more fulfilling life—look for the *how* rather than the *why*.

Get in touch with your feelings

One way to enhance your psychological well-being, and fortify your switch craft, is to learn to get in touch with your feelings. It's natural to try to suppress our feelings as a way to regulate our emotions, but it's often not very effective. On many occasions, learning to experience and accept our feelings can be a powerful way to regulate stress and anxiety.

Expressive writing is a surprisingly effective technique to achieve this. One study involved a group of computer engineers who, having been made redundant after twenty years working for the same company, wrote down their deepest thoughts and feelings about losing their jobs and how their personal and professional lives had been affected. Another group, drawn from the same pool of newly redundant staff, were asked simply to write about their plans for that day and their job-search activities, without thinking about their feelings.

The results were eye-opening. Not only did the people in the expressive writing group feel much better and experience fewer mental health problems, but almost half of them had also found a new job within eight months, compared to only a fifth of those in the control writing group. Learning to embrace your emotions by simply writing down how you feel about different situations can help to rewire your approach to life. But a word of warning. Studies show that expressive writing, especially about upsetting situations, can be very uncomfortable at the beginning. After about two weeks, however, these early costs fade, and the benefits begin to emerge. So stick with it, because the benefits of expressive writing are long-lasting and will have wide-ranging positive effects on your well-being.

There is a fine line, of course, between embracing your negative emotions and wallowing in them. There are no hard-and-fast rules, but most of us over time will learn to know the difference. One study has shown that the key to psychological health is neither to bury our negative emotions, nor to dwell on them for too long either. Across several separate studies, this research found that embracing negative emotions did not lead to an increase in wallowing. In fact, the direct opposite happened: accepting negative feelings led to *less* ruminating about them—the negative feelings lifted once their job had been done. Remember that emotions are naturally short-lived— it is thinking and ruminating about the bad feelings that results in an ongoing state of wallowing or continuous bad feelings, as seen all too clearly in Alan's experience. We are much more likely to thrive when we treat negative feelings like passing clouds.

Learning to push your boundaries

Once you become more familiar with your emotions you can then start to look for ways in which you can stretch yourself. I worked with a recently promoted senior executive, Andrea, who was nervous about public speaking. She had coped with this over many years by either bottling up her feelings or avoidance. Neither is a very effective strategy for this particular situation. But now, in her new role, Andrea could no longer avoid presenting to larger groups, and realized that she needed to take control of her fear. What works best in situations like these is a series of minor tweaks; a gradated exposure so to speak. So we worked together on identifying the level at which Andrea was comfortable. She was fine in meetings of up to about six people, but what really petrified her was the thought of standing up and speaking to a group of sixty people. "What about forty people?" I asked. "Terrifying," she responded. "What about twenty, or ten?" I continued. "I suppose ten would be OK," she agreed.

I gathered twelve students together and asked Andrea to present a short talk to them. The students were engaged and, to her surprise, Andrea really enjoyed the experience. After a few more practice runs, she began to grow in confidence and came to realize that it was not as difficult as she had thought. By pushing her limits in a way that was not too overwhelming, she had learned to regulate her fear.

Facing up to your fears in a safe environment will help you to progress. Bottling your feelings or avoiding challenging situations won't. The key is to set yourself short-term goals that stretch you but do not overwhelm.

Learning to distance ourselves from unhelpful thoughts or feelings

An analogy that I often use—both with the students I teach and the clients I consult with—is that our comfort zones are like walking round a wildlife park. All the animals—our "dangerous" or "uncomfortable" emotions—are safely ensconced in their enclosures and we can observe them, in safety, from a distance. Open the doors to those enclosures, however, and it's a different story: panic!

Of course, what would make such a scenario far less daunting is if we were trained and knowledgeable in handling these "wild" emotions, much like Andrea eventually learned to "tame" her anxiety over public speaking. There are plenty of techniques to choose from. One, for example, comes from mindfulness practice and is designed to help you find ways to create a distance between yourself and your unhelpful thoughts so that you can allow them to come and go without a struggle.

The "practice of diffusion" is a good case in point. The idea is to write down a couple of self-judgmental, negative thoughts: "I'm fat" or "I'm boring" or whatever resonates with you. Now pick one of them and spend about thirty seconds giving it your full attention, trying to believe completely in the thought as much as you can. Next, replay the thought with the phrase "I'm having the thought that . . ." added on to the front. For instance, you might say, "I'm having the thought that I'm boring." Once you've done this for several seconds, then add the phrase "I notice I'm having the thought that . . ." in front of the original. Now practice this over and over with different negative phrases, until you begin to see that it is not that difficult to distance yourself from them. Distancing is a powerful mental tool that will show you how thoughts can come and go, and that they are not necessarily linked to reality.

The worry chair

Another technique that I particularly like is the "worry chair." Worry can fuel negative feelings and so a good way to regulate the bad feelings is to find ways to manage the worry. Go to a quiet room and place a chair in a clear spot. Sit in it and connect with a negative thought or worry that has been playing on your mind. You might be worrying that a performance review next week is not going to go well, for instance. Again, take a minute or two to fuse with your thoughts around this negative belief, really imagining that the review has gone badly. Perhaps your boss is concerned and has given you a probationary period to improve. Notice how trapped and miserable you feel. Allow yourself to become immersed in how it feels. When you have had enough of this, move over to the other side of the room and look and observe your worrying self as if you're still sitting in the chair. Notice how miserable you look and identify either with your "worrying self," back in the chair, or with your "observing self" that is here, merely watching the other.

This technique is another powerful mental tool that can help to separate yourself from your worries and their associated negative feelings. Gradually you will learn to move easily between the two perspectives. Somebody consumed by grief following the death of their partner may learn to distance themselves from their "grieving self," but on certain occasions such as an anniversary, they will still be able to identify with their grieving self and remember their loved one.

Self-negotiation therapy

A third technique—sort of related to the "worry chair"—that I have adapted from the field of conflict resolution and use a lot in my coaching is what I call "self-negotiation therapy." Most hostage negotiators follow a five-step approach to get someone else to see your point of view and change their behavior. While you may never have to talk someone contemplating suicide down from a ledge, or deal with an armed stranger, these steps are just as useful to work through in everyday situations:

- **Active listening:** Take the time to genuinely listen to what someone has to say, without interrupting.
- **Empathy:** Try to understand why they are feeling the way they are.
- **Rapport:** Use your social skills to establish some rapport, perhaps by using humor, or telling them about a time that you have felt the same way.
- **Influence:** Once you have some rapport and have a deeper understanding of what's going on, perhaps very gently try to persuade the person to take a few deep breaths and talk further.
- **Behavioral change:** Hopefully at that point the person will begin to change their behavior and their plans.

Most people naturally try to skip the first three and go straight to trying to solve the problem, especially when the situation is pressured. However, skilled negotiators will tell you that this won't work, because listening is the most crucial step. It's only by listening in a very genuine way that you begin to understand where the other person is coming from and how they are feeling. Once you have this empathy, you can then build rapport, where the other person begins to trust you. It's only then that you can begin to try to solve the problem together. At every step in this process a good negotiator must be adaptable and flexible, empathizing with the person, calming the situation, and all the while keeping their own emotions under control.

Remember that someone in a crisis does not want to listen, they want to talk. Think of the last time you were really angry: were you interested in listening to what the other person was saying? Thought not. When we are in a crisis we are often guided by overwhelming emotions and not acting rationally. In this situation it's critical to listen, really listen, to what someone has to say. Genuine listening will eventually defuse the strong emotions. Really listening to people and building empathy and rapport can prevent the escalation of future "hot" emotions escalating. Once again, these skills are useful parts of the switch craft practitioner's tool kit.

When asked to list the top attributes of a good negotiator, experienced police crisis negotiators gave the following top three answers:

1. An effective listener
2. Patient, calm, and stable
3. Flexible, adaptable, and a quick thinker

Nearly half highlighted "maintaining flexibility and thinking on your feet" as being vital for building rapport. These same skills are essential to build our switch craft in noncrisis situations. By learning to regulate our own and others, emotions we can learn to better manage the ups and downs of everyday life.

The thing to remember is that good negotiators always have a plan; they never just "wing it." While they are listening carefully and building rapport, they will always have an end goal in mind and everything they do is designed to reach that goal, while still using a flexible approach. Like so many situations in life, including switch craft, the important thing is to understand what works, why it works, and then to practice and practice and practice until you can act or think in that way very easily and fluidly.

It is important to put this habit of good planning into operation in your own life; it is essential to boost the third pillar of switch craft. But—and here's the rub—not just when you are anticipating a difficult conversation, or difficult situation, with another person . . . but also when you're anticipating a difficult conversation, or experiencing a difficult situation, with *yourself*.

1. Actively listen to *yourself*.
2. Empathize with *yourself*.
3. Build rapport with *yourself*.
4. Influence *yourself*.
5. Change *your* behavior.

Have a clear plan and decide what you will do following different potential outcomes. Having a plan like this will give you a sense of control and will help you manage your emotions in advance so that you can react more appropriately in the moment.

Labeling our emotions is important

Most people are aware of common emotions such as fear, anger, happiness, or disgust, but there are so many "complex" emotional experiences, such as awe, pride, jealousy, sensuousness, tenderness, or relief. And your ability to describe them has implications for your psychological well-being.

When you heard that Donald Trump had lost the 2020 US presidential election, were you overjoyed, elated, happy, or relieved? Perhaps you felt crushed, disappointed, shocked, or outraged? The capacity to describe emotional feelings in fine-grained detail is sometimes called "emotional granularity." Just as the Inuit people have many words for snow, some people have many words to describe their emotions and can draw detailed distinctions between feeling fearful, repulsed, angry, sad, or apprehensive. However, those with low emotional granularity may not be able to describe unpleasant feelings in any more detail than just "unpleasant" or "bad." Being able to more precisely label your feelings brings them into higher definition and helps you interpret what they mean. It allows you to understand your trigger points—why you become anxious or apprehensive *about* something or why you become angry *with* somebody. Ultimately, having the words to explain how you are feeling with great clarity allows you to calibrate yourself to manage a variety of circumstances. It's why these emotion regulation skills are useful for switch craft.

The power of emotional granularity to improve our emotion regulation skills has been shown in the following study that teaches us how we can learn these skills ourselves. Volunteers were asked to indicate the extent to which they had engaged in different emotion regulation strategies over the previous two weeks. These included things like reframing by trying to find the silver lining in difficult situations, distracting themselves by withdrawing from an unpleasant encounter, or trying to actively engage in enjoyable activities and so on. Then, people were given daily diaries to keep for a fourteen-day period so they could rate their most intense emotional experience each day. Examples of positive emotions they might note down included joy, happiness, enthusiasm, and amusement, while negative emotions included nervousness, anger, sadness, and shame. What the research showed was that people who found it difficult to describe ways of distinguishing between their negative emotions were less likely to use effective strategies to regulate those emotions. Those who were able to label their negative emotions in a fine-grained way used many more strategies to cope with adverse life events.

Labeling our positive feelings in more fine-grained detail has also been found to be beneficial and has been linked with resilience. The good news is that emotional granularity can be learned. So, next time you feel strong emotions—whether positive or negative—try to find a variety of words, perhaps even from different languages, to describe them. Learning a variety of words to simply label your feelings is one of the more surprising and simple ways you can learn to regulate your emotions and improve your psychological health.

Chapter Summary

- Regulating emotions is an important part of switch craft because it allows us to step back and assess a situation clearly so we are in a better position to consider the most appropriate approach. Strong emotions can push us to act on impulse, which might not always be the right solution, so learning to regulate them is an essential part of the switch craft tool kit.

- When managing our emotions, we need to strive for agility and authenticity in equal measure. It's great to be able to reframe things positively, but being falsely positive can be detrimental.
- There are several different ways to regulate our emotions, and flexibility choosing between them as situations unfold is crucial. There is no one-size-fits-all solution.
- One possible exception is learning to label your feelings, good and bad, in fine-grained detail. The greater "emotional granularity" this gives us has been shown to be a good way to help us regulate emotions and enhance well-being and resilience.

THE FOURTH PILLAR OF SWITCH CRAFT

Situational Awareness

CHAPTER 12

THE NATURE OF INTUITION

Back in 1984 when I was a student, I spent a happy summer working in restaurants and having fun in the United States. It was a long hot summer and my friend Maria and I landed on our feet when we got jobs as chambermaids in a Montauk hotel, right at the tail end of Long Island in New York. The job wasn't very well paid but came with lodging for us both right at the edge of a long sandy beach. Work started early, around 6am, and ended just before lunchtime. We then had the entire afternoon to spend on the beach before starting our second jobs working in an expensive fish restaurant in the evening.

The large tips rolled in and with no accommodation costs and lots of free time we were having a blast. Montauk was New York's beachside playground, packed with people holidaying from Manhattan and further afield. We quickly met lots of other students from all around the world working in the numerous bars and restaurants around the resort.

We became firm friends with an American girl, Jenny, who was traveling on her own and was working in the same hotel. She was vivacious and full of fun and introduced us to many people she knew. None of us were surprised when she attracted a handsome boyfriend; I remember vividly how envious the rest of us were. We saw Jenny less and less as she spent more and more time with him.

One afternoon, I went down to the beach on my own and bumped into Jenny and her new boyfriend. I was struck again by how good-looking and charming he was. But I was also struck by something else. It was very subtle, but I quickly became aware of a vague sense of being uncomfortable and apprehensive. I couldn't pinpoint what it was, but as I spoke to him his gaze lingered just a little too long. It was not at all in a flirtatious way, it was slightly hostile, and unsettling. On subsequent occasions over the next few weeks I got the same gut feeling and became increasingly uncomfortable near him. I remember being surprised because he had not actually done or said anything that was intimidating. And yet, I had a persistent sense of unease and became quite wary of him. I told Maria about it one evening and it turned out that she also had an uncomfortable feeling about him.

A few weeks later Maria and I were woken in the middle of the night by a loud banging. It was Jenny's boyfriend kicking and punching our flimsy door, demanding to know where she was. He was incandescent with rage and convinced that she was hiding in our room. We were able to open the door but keep the screen door locked shut so that he could see that Jenny was not in our room. But his fury did not diminish, and it was terrifying. He punched through the screen, shouting and insisting that we must know where she was, before he finally stormed off.

Thankfully, Jenny was shaken but fine. She told us she had become increasingly scared of him in the previous weeks. In the next few days, we were all interviewed by the police with probing questions of what we knew about him and whether we knew where he might have gone. We were shocked to discover that he was on the run and wanted for several rapes committed in California earlier in the summer.

It was a chilling experience and, to this day, I couldn't explain to you why I felt danger and apprehension in his presence despite all the "external" evidence that he was a charismatic, good guy. But my brain was clearly picking up cues that gave me a warning signal. Many of our "intuitions" are based on an intimate understanding of our surroundings, the context, and this is a vital skill that we can develop and hone. Intuition is an important part of switch craft for exactly this reason—by providing us with often subtle information from our surroundings, intuition helps us to make the right decision at the right time.

The nature of intuition

Many studies in psychology tell us that intuition is a very real process where the brain makes use of past experiences, along with internal signals and cues from the environment, to help us make a decision. This decision happens so quickly that it doesn't register with our conscious mind. This was demonstrated in a now classic study in which volunteers had to choose cards from two decks, which unknown to them were rigged. One deck gave big wins and big losses while the other gave small gains and almost no losses. It took almost eighty goes on average before the volunteers figured this out. But here's the interesting finding: after about only ten cards, volunteers had a sense of which was the "dangerous" deck; upon further investigation, the researchers found that the volunteers experienced what is known as a galvanic skin response—increased sweatiness—when choosing the high-risk/high-gain cards. The researchers concluded that this bodily signal generated an intuitive bias that was used to guide decision-making before the conscious brain was aware of what was happening. It's clear that in many situations where we must make decisions without having all the facts, having access to this intuitive sense can be advantageous.

Intuition is that part of our mind that presents us with the *gist* of a situation. These intuitive hints are almost imperceptible, occur rapidly, and allow us to pick up information about the world without intention. It is knowledge that has an unarticulated feel, not taught but acquired by osmosis. Intuition provides the bedrock of our ability to understand complex everyday situations and problems. It's the "gut feeling" that's easily missed. Most of us will have had that gut feeling that something is just not quite right, even if we cannot pinpoint why. It's what I experienced when I had an unsettling feeling about Jenny's boyfriend. These intuitions can also be very helpful in tuning us in to the cultural norms of a new situation. Because our brain analyzes patterns and probabilities before our conscious mind has time to catch up, these intuitive hints can be very useful to orient us when we are in new and unfamiliar surroundings.

For instance, when I completed my PhD studies at the tender age of twenty-five, I was beyond excited to have landed my first academic job in New Zealand—a country that seemed a million miles away from Dublin. Although Ireland and New Zealand had broadly similar cultures and language, moving was a culture shock. The following weeks and months were a steep learning curve. I learned quickly not to make frivolous jokes about rugby. What was a much-loved sport back home in Ireland turned out to be a national religion in New Zealand and taken very seriously.

No one told me this directly; but I learned fast. Observing people's reactions and how they spoke among themselves about the game told me all I needed to know. This type of intuitive knowledge is difficult to explain; we just *know* when something is right and when it is wrong. It is knowledge that guides our behavior without necessarily being available to our conscious awareness. Deeply intuitive people are often thought to have mysterious powers that they can draw upon from the universe, a spiritual source, or some internal hard-wired part of their brain. They seem different in some fundamental way from the rest of us.

Intuition is not magical

The truth is, however, that intuition is not magical. Instead, it is an extension of how your memory and cognitive systems normally work—a mental skill that is affected strongly by your life experience.

The way it works is that your brain gathers as much information as possible, checks this information with the "big data" from your prior experiences, and then makes a prediction. Catching a glimpse of a close friend entering a shop in bad light as you swish past in a car leads to instant recognition. Your brain does not have enough information to make a rational and detailed identification but there are sufficient cues in the shape of her face, the way she walks, or the sweep of her hair that allow you to make a rapid decision.

This ability to deduce vital information based on only narrow slivers of experience has been called "thin slicing." It has been studied intensively by looking at the profound impact that first impressions can have. In one well-known study, students were asked to evaluate their professors at the beginning of their first year before any classes had begun from a ten-second video clip and then again two years later after multiple classes and interactions. The two evaluations were almost identical. This shows us that the initial instinctive impression remained in place and did not shift over a long period. First impressions really do count, even though they may not always be right.

Intuition comes from experience

It's important to note that intuition is an elusive form of intelligence that we acquire through personal experience rather than by means of deliberate learning. Next time you are at a computer, try to type the sentence "Jack was a long way from home" without looking down at your hands. My guess is you will find this fairly easy; but if I then asked you to name, without looking, the ten letters that appear in the middle row of your keyboard, you would struggle. Having to recall the location of the letters relies on explicit memory, which is acquired deliberately, whereas typing relies on your *intuitive* memory, which is typically acquired unintentionally.

Much of our everyday competence is based on skills and information like this that we have picked up without any explicit instruction.

Think of the many social customs that we understand in an unspoken way, or the way we acquire language as young children. Even with little formal education, native speakers of a language have an intuitive grasp of grammar but often cannot explain the rules of that grammar in any detail. It is based on what is called "tacit knowledge"— knowledge that we know we have, such as tying a knot, riding a bike, or catching a ball, but cannot easily put into words. It is typically learned through actions and everyday experience rather than books or lessons. And we only know we have this knowledge when we perform it, whether it's a complex dance move or our quick reaction to a dog running into the road when we're driving. Donald Rumsfeld, the former US Secretary of Defense, famously spoke of the "known unknowns"—the things we know we don't know, but there are also things we know but don't know *how* we know.

Intuition tells us what's important

To put it another way, intuition provides us with an instinctive understanding of what's important and what information can safely be ignored. Remember when you were learning to ride a bicycle how you paid intense attention to every move you had to make until gradually the complex series of movements became automatic? As we become expert with a new skill we gradually pay less and less attention to every detail. In fact, the development of expertise involves precisely this: learning to attend to what's most important and letting the brain deal unconsciously with the rest.

Intuition guides our perceptions

First and foremost, intuitive powers help with our survival; intuition can guide us toward the most relevant aspects of a scene even when we don't know it. My own work on the profound impact of danger signals on our attention—perhaps inspired by my formative experience with Jenny's boyfriend in Montauk all those years ago—provides a good example of how our attention orients rapidly to a danger cue: an angry face, say, in a crowd of neutral expressions. No real surprise there. In one study I conducted, volunteers watched as a series of images were flashed up in different locations on a computer screen. Some of these images portrayed angry or fearful facial expressions while others were more pleasant, with happy smiling faces. When we measured where their eyes and their attention tended to land what we found was that the angry expressions drew their attention much more than other expressions. What was surprising was that when I prevented the conscious perception of the faces by presenting another jumbled image over them after just seventeen milliseconds— blocking the perception of the faces—the now invisible danger signals had an even stronger draw on attention. Even though my volunteers were not aware of what, if anything, had been presented, the angry faces still pulled their attention much more than the happy faces. Further studies showed us that a galvanic skin response was alerting my volunteers to the danger cues, just as in the earlier studies with different decks of cards.

This is an example of intuition—or what's often called a "gut feeling"—in action. A subtle bodily cue pushed volunteers to pay more attention to these images. Our gut feelings are important and ensure that we pay attention to the most relevant aspects of our environment. It's important to remember, however, that gut feelings are not always correct. That's not the point of them. Gut feelings will not give you clear right or wrong answers that you can then rationally evaluate. That is simply not their function. You may not always get it right, as intuition is not a black-and-white art. Nevertheless, your intuition gives you bonus evidence to guide your judgment.

Albert Einstein has been widely quoted as saying, "The intuitive mind is a sacred gift, and the rational mind is a faithful servant. We have created a society that honors the servant and has forgotten the gift." Gut feelings, or intuitions, are there to *guide* you toward a more rational analysis and help you to adapt in dynamic and rapidly changing environments. There is little doubt that we need both— intuitive perceptions *and* rational analysis—to make the best decisions we can. In the following chapter, we will explore some tips and exercises that can help you to tune in to your intuitions. Part of that is learning to quieten the chatter in your mind and your surroundings and pay more attention to your bodily signals. First, let's have a closer look at why this works.

Intuition really is based on "gut feelings"

To gain a deeper understanding of intuition, scientists are beginning to turn toward the gut and how it operates. It turns out that the term "gut feelings" is surprisingly accurate. These intuitive signals do indeed come from a layer of neurons lining the stomach and gastro- intestinal tract that is often referred to as the "second brain." Called the "enteric nervous system," these gut-based neurons are intimately connected with the brain and help us to turn environmental signals, such as danger cues, into vague feelings of danger that we can then act upon. There is much we don't know about gut-brain interactions, but there is little doubt that joined-up thinking between the brain and the gut helps us to navigate a fast-paced world.

The importance of context

What's important to remember is that information coming from our intuition is not stand-alone information. Our gut feelings *inform* rather than dictate. When you are at the cutting edge, whether it is in science or in business, there is no guide. You have to step out fearlessly into the unknown, and that is precisely when your intui- tion comes into its own. Intuition is to context what a surfboard is to a wave.

Looking outside allows us to recognize the ambience of a situation, no matter how nuanced, and instinctively understand what's important and what we should do. Originating from the Latin word *contextere*, the meaning of "context" was originally "to weave together" or "to interlace" the meanings contained within a text. Nowadays we use the word in a much wider sense to refer to all those circumstances that influence how you feel and behave in different situations. It might be your culture; it might be a fleeting memory of a similar situation in the past; it might be the presence of a particular person. Our immediate surroundings play a big part in determining the roles we play and how we feel, and is known as the "performance context." You might, for instance, think and feel very differently about certain things depending on whether you are at work or at home.

I had the opportunity to test this when one of my students was working on a project to investigate the experience of employees in a company and wanted to conduct some short interviews. Approaching people at the end of a long and busy day as they were leaving work, we realized, was likely to elicit completely different answers than if the interviews took place in the office during a morning coffee break. We then wondered if we would get the same answers if the questions were asked in the workers, home environment rather than at work? As expected, each different context resulted in a different quality of answers.

Context is an important factor in determining achievement and success. Of course, in a wider sense context becomes culture and people bring their culture, traditions, and economic reality to every situation—as I did when I first moved to New Zealand. All of this has a profound influence on how we respond in any situation. In a series of studies conducted in rural Kenya, researchers wanted to discover whether "intelligence" was a universal concept. Adults were asked to assess the "intelligence" of the children in their village; the children they rated most highly were those who had learned how to apply various herbal medicines. This made sense, as parasitic infections are common in these rural villages and only a few of the hundreds of available herbal medicines are effective in helping with the resulting stomachaches. The children who had learned how to find the best herbs to self-medicate had an adaptive advantage. Interestingly, these adaptive children tended to do *less* well on conventional Western school tests. The authors concluded that school achievement had a very low value in these villages where most of the children do not complete high school. In fact, children who stayed on and excelled at school were often seen as wasting their time, as those skills would not help them to find a job and become economically secure. "Achievement" can be understood only within a cultural context. A Kenyan child who does not do well in school is no less intelligent than an American or European child who does not know much about herbal medicines.

Our ability to excel at what is most valued in our immediate context is the best predictor of our ability to adapt and flourish. Many African and Asian societies place a much higher value on social qualities like respecting and caring for others, diligence, consideration, and ability to cooperate than on Western-based notions of success. For the Baoulé people of West Africa, respect for your elders and service to the community is seen as the cornerstone of intelligence. There, the ability to facilitate stable and happy intergroup relations is considered to be far more important than the ability to solve problems that is given higher importance in many Western countries.

Not that solving problems *isn't* important. Alan Sugar, the host of the TV show *The Apprentice* in the UK, once commented that he was looking for an apprentice who was "drop-dead shrewd"—someone with street smarts rather than book smarts. The two don't always go hand in hand. That is because "academic" problems often don't represent the type of problem that faces us in everyday life. Everyday problems are more meaningful to us and often have many different possible solutions, each with many pros and cons.

Intuition is based on the richness of our experience

Intuition is fine-tuned by the diversity of different contexts that we have been exposed to; it's why people can be very intuitive in one area of expertise, but this does not necessarily transfer neatly to other domains. This suggests that all of us can improve our intuitive powers if we put our mind to it.

Experts in a specific domain, whether it is nursing, computer programming, or leadership, develop a high degree of intuition that comes from years of experience within a specific field. What's important is the diversity of this experience. Take nursing. A nurse does not do just one thing over and over. An experienced nurse will have worked in many different contexts, will have witnessed people die, seen people survive, watched people react to both good and bad news, and worked in many different subspecialties. There will have been many occasions when she or he may well have had to "think outside the box" and use what was to hand in an emergency. This diversity—within nursing—leads to a deep and intuitive understanding of most situations that are likely to arise within a specific professional domain.

We can also see this demonstrated in the work of those people who decide on how much weight each racehorse must carry in a race. This is known as "handicapping," where horses are allocated different loads to try to even out the chances of them winning. The horses who are likely to perform best carry the heaviest loads, whereas those likely to perform less well are given a lighter load. Working out how to best handicap each horse is a mathematically complex process that takes into account myriad issues that determine how well a horse might run on a particular day. Many factors make a difference: previous results, the impact of different weather conditions, whether the horse is comfortable passing other horses, whether attempted passes have taken place in the past, and so on. Expert handicappers use a complex algorithm to predict the likely speed of each horse and to work out the odds of each one winning.

You might think that a person's ability to use these complex algo-rithms would translate into other areas of skill. However, in one study researchers found it was unrelated to IQ. In fact, one of the most successful handicappers was a construction worker with a low IQ of 85. The researchers then asked the expert handicappers to have a go at making stock-market predictions, which use very similar algorithms. Even though the calculations required were very similar, the different context with which they had no familiarity resulted in performance that was no better than chance. They might as well have been guessing.

The context can act as a cue to behave in a certain way; take away the context and the behavior changes. Our intuition is the result of accumulated knowledge that, while hidden, is especially vital when we need to make quick decisions under pressure. When there is not enough time to absorb all the information necessary, our intuitive intelligence steps into the breach and comes to our rescue. It's not always right, of course—as mentioned previously, that's not it's job. But it guides us, using evidence from our own past experience: the more diverse this experience, the more helpful and constructive our intuition. Intuitive people have developed the practical ability to learn from experience; they can pick up on even the subtlest of cues and are able to turn this knowledge to their advantage.

The value of intuition is often downplayed in the curriculum of business schools, even though senior executives have been shown to frequently rely on intuitions as well as critical analysis to make impor-tant decisions and achieve commercial success. In the business world, the climate is often complex and unpredictable, making traditional logic-based decision-making ineffective. This is where switching to intuition and instinct can guide a businessperson to a vital piece of information that might have eluded our rational mind. This can give a competitive advantage, especially when steering a business through uncertain times. For instance, businesswoman and entrepreneur Estée Lauder, who founded the cosmetic company, was known for her uncanny ability to "out-predict" detailed market research and identify which fragrance would sell. Some of her employees speculated this was because of some supernatural sense, but more likely it was down to an intuitive ability to understand people and their deepest desires.

Like most forms of intelligence, intuition is not static but develops with experience. According to her colleagues, Estée Lauder would spend hours and hours talking to her clients, learning about what they liked. It was this intimate knowledge that allowed her to develop a look and a feel that she knew, intuitively, would appeal. In the same way, through experience, physicians can develop a rapid sense of a complex diagnosis within minutes of talking to a patient, while soldiers with many years in the field can intuitively sense danger without being able to explain why. And in everyday life, because of prolonged experience, we can tell within the first few seconds of a phone call when our partner is angry, or quickly recognize when our child is concealing some wrongdoing.

The cognitive scientist Herbert Simon explained this well when he said, "The situation has provided a cue; this cue has given the expert access to information within memory, and the information provides the answer. Intuition is nothing more and nothing less than recognition." Intuition develops when familiar things are recognized, and acted upon, in a new situation and this only comes from experience. Perception works in the same way; minimal information helps us to recognize and predict familiar elements even when we think it is impossible.

Given that switch craft is all about understanding a situation as accurately as possible, acknowledging the power of intuition is important. If we ignore what our "gut feeling" is telling us, we may well make poorer decisions in many walks of life.

Chapter Summary

- Intuition provides us with the gist of a situation and is a useful guide for decision-making.
- Our context provides much of the data that our intuition uses to produce "gut feelings."

- Intuition cannot be taught; but it can be enhanced by exposing ourselves to lots of experiences.
- Intuitions are not always correct; instead, they are bits of additional information that can be used to guide our more rational decision-making.
- Intuition is relevant to switch craft because it provides us access to a level of knowledge about the world that is not available to our conscious mind. This type of knowledge is often most important in complex, fast-changing situations—exactly where switch craft is most needed.

CHAPTER 13

LOOKING OUTSIDE:
HOW CONTEXT FUELS INTUITION

In the fast-paced world of Formula One motor racing, forty-year-old Argentinean driver Juan Manuel Fangio was one of the best. In the 1950 Monaco Grand Prix, Fangio was on his second lap and well ahead of the pack as he entered a tunnel with an infamous hairpin bend. He usually accelerated as he emerged from the tunnel to maintain his speed, but this time, inexplicably, he took his foot off the pedal as he exited and slowed down dramatically. It was lucky he did. Just around the corner was a horrendous crash with a pileup of nine cars strewn across the racetrack. If Fangio had not slowed down, he would have ploughed straight into them. Instead, he was able to find his way through the debris and go on to win.

After the race, Fangio wasn't initially able to explain his sudden intuition to decelerate. But eventually his team uncovered the cue that Fangio had noticed. As he was in the lead, he would normally have seen a sea of mostly pink faces looking toward him from the stands as he emerged from the tunnel. But instead he noticed a darker blur of the backs of spectators, heads, as they had all turned to look at the crash site. This subtle change of shading had registered in Fangio's brain, told him something was wrong, and caused him to slow down. Based on his extensive racing experience, Fangio was able to recognize an unusual pattern of cues in his environment (heads of spectators turned the wrong way), which caused him to react intuitively in microseconds.

A vital component of intuition is the capacity to look *outside* and better understand what is going on around us. As we have seen in the previous chapter, cues from our environment trigger past knowledge that informs the current moment and what becomes available to our conscious awareness is a general sense that something is not quite right. A vague feeling that he should slow down was triggered by Fangio's brain picking up the unusual cue of spectators looking the other way. These subtle feelings provide us with inside information from our brain's vast reservoir of past experience. While the information is not always correct, it does guide us in critical moments where full rational analysis is simply not possible. This is intuition, as we explored in the previous chapter, but intuition driven by a supreme sensitivity to the context. Called "context sensitivity" in psychology, this is a field of research that has been surprisingly neglected. It refers to the capacity to figure out what is needed in a particular context. Experts will be alert to their own context, as we saw with Juan Manuel Fangio, but for switch craft it is helpful to be sensitive to as wide a range of contexts as possible, especially social ones, because this provides us with vital ingredients to nurture our switch craft skills. In this chapter, we will explore ways in which we can improve how sensitive we are to our surroundings and how, in turn, we can improve our intuitive powers.

Context sensitivity in action

I recently got a personal taste of context sensitivity in action when I paid a visit to a psychic. In the interests of science, of course. I was a little nervous as I walked up a couple of flights of stairs to an unexpectedly bright and airy suite of offices in north London. I sat with Anna at a small, highly polished desk near the window, not a crystal ball in sight. "How can I help?" she asked. But before I had a chance to answer, she squinted at me and said, "I can see you are struggling with a major decision." She predicted that I would make the right choice soon and find an exciting way forward. This was, of course, a somewhat vague observation but it *was* true that I had recently been offered an exciting opportunity outside academia and was seriously considering a radical change in career direction.

I was impressed and started to relax in Anna's company. When she dealt my tarot cards, they all confirmed Anna's initial prediction. She first turned over the Death card, which alarmed me, but Anna reassured me that this meant one major phase of my life was coming to an end and a new one was about to start. She could not be sure of the timescale, but it would not be too long. She told me that I was stressed about the change and that the card was telling me to take the plunge and act now. She turned over several other cards but glossed over them until she turned over the Wheel of Fortune. "Ah," she said, "this is really interesting." She explained that this card was telling me that a big change was going to occur within the next year and I should get myself ready to adapt.

I left the session feeling strangely relaxed, with the feeling that she had "seen" elements of what was going on in my life and left me with the positive vision that things would work out really well. Whatever my preconceived view on psychics, Anna had accurately picked up on the mental turmoil had I been grappling with. I realized that she was probably very skilled at picking up subtle signals from me when something resonated. She was what is called in the trade a good "cold-reader," meaning picking up subtle cues when something that is said is important to someone even when they are trying to disguise it. She was very "context-sensitive." Statements like "I can see you are struggling with a major decision" are very effective because they feel personal but are actually very general and can relate to anyone. When you hear something that is pertinent it's hard not to react. A good cold-reader will pick up on this reaction and then develop the theme. Even as a professor of psychology trying to present my best poker face, I had failed to conceal my personal connection with Anna's generic statement. I almost certainly "leaked" cues telling her that she had struck a chord.

How can we improve our context sensitivity?

So, how can we develop our context sensitivity—and, by default, our intuition? More importantly, why should we do so? The answer to these questions is clear. The answer to "how" is simply to "get out more" (in a manner of psychological speaking). And the why? Because it will enhance our switch craft skills, help us to make smarter decisions, as well as enhance our resilience and general well-being. So, plenty of good reasons. Switch craft requires us to be tuned in to our environment, and "context sensitivity" is a crucial part of this.

Experience is everything

Anna told me she'd been working as a psychic for over thirty years. She'd originally set up as a medium in Blackpool as a teenager, before working the festivals and then moving to London. I wasn't in the least bit surprised, given how good she was at cold-reading. We acquire intuitive knowledge not by intellectual analysis, but by *doing*. Reading manuals about how a radio works will tell you a lot, but to fully appreciate the inner mechanisms there is no substitute for taking it apart into a thousand pieces and physically putting it all back together again. This type of hands-on experience is transformative.

Doing rather than thinking is the way to help your brain build its reservoirs of intuitive knowledge. Exposing yourself to many different experiences—and their attendant thoughts and feelings—will help you to build your sensitivity to the context, which provides the scaffolding for your intuition. Experience is difficult to undo. Once we have seen, we cannot easily unsee, and once we have understood, we cannot easily misunderstand.

The importance of "getting out more" is captured in a quotation that is commonly attributed to C. S. Lewis, "Experience: that most brutal of teachers. But you learn, my God, do you learn." We all need to develop a useful experience bank that we can draw upon when needed. To achieve this, it's essential to immerse yourself completely into daily life, warts and all. You can go on retreats, study in libraries, debate endlessly with your friends, but you will only truly learn by exposing yourself to genuinely new and different experiences, by confronting yourself with people whose views you disagree with, and by pushing yourself well outside your comfort zone. As C. S. Lewis himself reminds us: "It is much easier to pray for a bore than to go visit him."

Diversity of experience helps our switch craft because it shifts the internal learning algorithms in our brain to provide a more accurate interpretation of the context. Like a good children's story, life is laced with multiple layers of meaning and the wider your experience of life, the more likely you are to pick up on all the wonders and complexity of humans and the complicated situations they get themselves into. The good news is that a wide life experience will give you more cognitive, emotional, and behavioral processes to help you adapt to virtually any scenario with which you might be faced—so it is obviously a crucial component of switch craft.

Just as greater biodiversity is linked with greater resilience within an ecosystem, mental diversity is associated with numerous benefits to your psychological and physical well-being. Not only does mental diversity allow you to tap into a richer reservoir of intuition—a crucial pillar of switch craft—it also provides you with a wider variety of strategies and options to cope.

Work on developing your "situational awareness"

To improve our switch craft and really make use of our experience, it is essential to work on developing a broad awareness of the situation; this is the cornerstone of our sensitivity to context. Many soldiers who operated in the Iraq War, for instance, will tell you that a large part of the skill of survival was the ability to know what was "typical" and therefore be able to notice oddities—a heightened tension in the air, a surprisingly empty street for a particular time of day, and so on. In 2004 Lieutenant Donovan Campbell was leading a platoon of US Marines through the Iraqi city of Ramadi, on one of their regular road-sweep missions, to look for and clear bombs. They saw an IED in the middle of the road—and realized it was an obvious decoy. They were about to move on when one of the men noticed a concrete block about 100 meters away that he wasn't happy with because it looked "too symmetrical, too perfect." The block turned out to contain a massive bomb. "Unless you know what rubble in that part of Iraq looks like, there's no way you'd see that," said Donovan. It was the marine's knowledge of a *typical* street scene that allowed him to spot an anomaly. In a similar way, following the bombing at the 2013 Boston Marathon, police were able to quickly identify the bombers from CCTV footage as they calmly walked away while everyone else was panicking. Their odd behavior stood out.

A good understanding of what is "normal" in any situation allows you to rapidly pick out irregularities. Your brain is excellent at spotting anomalies so, counterintuitively, it's often better to look for consistency rather than for changes (or dangers), and then allow your brain to do the rest. Ask yourself, what is normal in any situation? In a café, for instance, you might expect people to be relaxed and comfortable, chatting to friends and drinking a coffee. People nervously looking around, or not talking to each other, may indicate that something's wrong. When you are in different situations, especially familiar ones, spend a bit of time noting typical behaviors, and then test your memory for them.

Sometimes narrowing your focus is essential

While stepping back and taking a broad overview of a situation is important, as we have just seen, it also pays to sometimes narrow your focus deliberately. A few years ago, my husband and I worked with Arsenal football club, to help find ways to enhance the cognitive skills of the players and improve their performance. We spoke with the players and coaching staff, and watched the players in action. Soccer is a fast paced and fluid game with many moving parts during most periods of play.

During intense training sessions, the challenge for a coach manager was how to keep his mind on all aspects of the performance when there were so many things happening. The answer, of course, is that you can't. It's an easy trap to try to stay across everything. Instead, the trick is to narrow your focus to just one aspect of the game, perhaps just two particular players, for a period of play, observing how well one is passing and how the other anticipates those intentions.

The trick is to try not to track too many different aspects of a situation as this will impair your ability to notice subtle aspects and therefore will undermine your switch craft. My research has shown, for instance, that we can keep a maximum of about four items in mind at any given time. This means that if you try to take in everything, especially in a new situation, your brain will quickly become overloaded. While this was found in experimental studies with simple objects, it's reasonable to assume that we should have no more than four items on our to-do list, as anything more will begin to exceed our mental capacity.

Paying attention to the eyes can help

Sometimes, as my experience with Anna demonstrates, intuitions can occur without us being aware of them and emanate from subtle aspects of body language that our brains are unconsciously picking up about others. And you don't have to be a psychic! Have you ever had the feeling that someone is lying to you but you just can't put your finger on why? The answer may well be in your context sensitivity to pupil dilation. Eyes may not actually be the window to the soul, but they are extremely revealing. A growing body of evidence tells us that pupil size is an important indicator of how someone is feeling. The underlying reason for this is because it is a reliable marker of mental effort. When you are thinking really hard, your pupils will dilate. You can test this out by standing in front of a mirror while you try to work out a difficult multiplication—say, 63 times 14—while looking carefully at your pupils.

Hiding deception also takes mental effort and this is why pupil dilation can be a real giveaway. In one study, some volunteers were asked to steal $20 from a secretary's purse when she left the room, while other participants did not steal anything. Later, all of the participants were asked to deny the theft. The researchers were much better able to detect the thieves when they analyzed pupil dilation. Those who lied about stealing the cash had pupils that were larger by about one millimeter when compared to the pupils of the honest participants.

It's not easy, but "cold-readers" like Anna become very skilled at noticing changes in pupil dilation. So pay attention to people's eyes, especially when they hear something surprising or exciting, and see whether you can begin to notice any changes. These can be good guides to an intuitive sense of how someone is really feeling.

Get out of your bubble

Easier said than done? Far from it. Most of us exist within bubbles of like-minded people, and not just on social media. So, look at those people you interact with most of the time. Are most of them people you work with? Are they mainly from your own family? Are they from a similar background and income level to you? Do they share similar opinions? Try to become more aware of this and take steps to meet a variety of people from different walks of life and really listen to them. Perhaps engage in different discussions on social media. Have a look at groups of people with very different political views and interests to you and see what they are thinking. Read a different newspaper or watch a different TV channel or volunteer to help out a charity in a different community from your own, get out to events that you would not normally go to. Anything that exposes you to cultures and people you are not familiar with will help you to open your mind and bolster your switch craft.

I tried this out when I was traveling in the United States in 2018. I was struggling to understand why so many Americans supported Donald Trump in the aftermath of what seemed to me to be a series of openly racist comments. This did not fit with the America, or the Americans, I knew. I made a point of listening to news channels that I knew were conservative and supportive of Trump's views. It was a real eye-opener to see how different media outlets conveyed the same news. I regularly watched one particular commentator with increasing fascination. I was astonished when I heard him apparently defending white supremacists and denigrating immigrants. But once I got over the shock of hearing views so different from my own, when I really tried to listen, I noticed that in interviews with guests, he sometimes made reasonable arguments. I even found myself agreeing with some of what he said. I began to understand the fears that many people have about the number of new immigrants coming into areas, especially areas where there is already high unemployment and poverty. Similar themes had emerged in the Brexit debate in the UK. While my fundamental beliefs on these issues are very different, I could see these views reflected genuine and deep concerns. Even a short-lived exercise like this can make you more open to different views and help your brain's predictions become a little less biased than they were before.

Using the right approach in the right context is crucial for switch craft

The reason why it is important to improve our sensitivity to the context is because it helps us to choose the right strategy for the occasion. We are only just beginning to understand the importance of choosing between different strategies to manage our emotions. As we saw back in Chapter 4 when taking about resilience, the flexibility of choosing between different coping approaches is essential. The same flexibility is required in terms of utilizing those psychological processes that help us to analyze the world around us. When you see a distant threat, like a faraway predator, say, the appropriate behavior might be to stay very vigilant and keep a close eye on what the creature is doing—but the same behavior when the predator is very close is not likely to end well. This is yet another example that every thought and feeling is only useful in the right context. As we know from switch craft, there is no one-size-fits-all for every situation.

Responding appropriately to the context is not only important for survival, but also a sign of good mental health. Those who thrive tend to choose the right approach for the occasion. For instance, studies show that children who are mentally healthy show strong fear responses to threatening situations, as you would expect, but *not* to low-threat situations—as you might also expect. In other words, their fear response is appropriate for the context. In marked contrast, children who show strong fear reactions even in low-threat situations—i.e. they don't show context sensitivity—are much more likely to develop subsequent anxiety and other mental health problems.

We see a similar inappropriate response to context in adults with severe depression. People who are not depressed typically show a strong emotional response to happy movies but a subdued response to sad films. In other words, they get a real boost from upbeat scenes but don't become overly down when watching sad scenes. In comparison, those who are depressed show a subdued response to both types of films. They don't get too down when watching sad scenes, but equally don't get much of a boost from happy scenes. Based on this type of evidence, it has been suggested that a lack of context sensitivity plays an important role in the maintenance of depression. The idea is that a type of "defensive disengagement" becomes an unconscious strategy in depression even when this is not at all helpful for the situation.

Why does experience improve our context sensitivity?

Exposing yourself to a wealth of experiences has a direct impact on how your brain operates. Remember that your brain is essentially a prediction machine that feeds on big data. Think for a moment about where this data comes from. It comes from all the sights, all the sounds, and all the experiences that bombard you from the moment you are born to the day you die. What this means is that your cultural and personal experiences have a profound impact on how your brain interprets and reacts to the things that happen to you. When many countries imposed strict lockdowns during the coronavirus pandemic, this inadvertently led to a much more restricted and narrow type of experience than usual. This is why many people struggled with social situations when they were allowed out again and, while this is a speculative theory, may well be why many people experienced "brain fog" after extensive social isolation.

Limiting our experience can also make us more vulnerable to biases. This can be seen in infants as young as three months old, who already show a strong preference for faces from the race/color they have been most exposed to during their short lives. In an interesting study, psychologists gathered three groups of three-month-old infants living in either Israel or Ethiopia and presented them with pairs of African (Ethiopian) and Caucasian faces side-by-side, then carefully monitored which of these faces the infants looked at more. This is a tried and tested method in psychology and if an infant consistently looks at one category of face more than another we can safely assume a preference. The results showed that the Ethiopian infants spent longer looking at Ethiopian faces while the Caucasian infants spent longer looking at Caucasian faces. They were most interested in what was familiar.

What is more intriguing is that Black Ethiopian infants living in an absorption center for new immigrants to Israel did not show this own-race advantage. Instead, they spent equal amounts of time looking at white and Black faces. The researchers were convinced that this was because of the diversity of faces that these infants were being exposed to on a regular basis.

A mountain of research suggests that our tendency to pay more attention to faces from our own ethnic group comes directly from the degree of exposure we have had to those faces early in life. Infants raised by a female caregiver show a preference for female faces, while those brought up by a male caregiver spend more time looking at male faces. A fascinating study shows that preference based on ethnicity is not present in the first days of life but rather is picked up within the first three months of life. While Caucasian three-month-old infants showed a preference to look at own-race faces, no preference was seen when the same experiment was done with infants who were just a few days old. The newborns were equally interested in faces from all races.

Our experiences provide our brain with big data

What happens is that our cumulative knowledge builds up an internal database of big data and this results in strong preferences; we orient toward that which is most familiar. The flip side of this process is the development of "dislikes" or reduced preference, for those from groups to which we have not been exposed to the same extent. In-group and out-group biases are among the most fundamental of biases, and the roots of these biases in the brain come from learning and exposure rather than from any hardwired biological "fear of the stranger."

In what has been called a "categorization instinct," we cannot help but group people into "us" and "them." This "us-them" distinction is often based on racial features, but it can also be based on nationality, people from your local area, supporters of your own sports team, or even members of your own family. This unavoidable instinct to categorize "us" versus "them" paves the way for the development of vastly different levels of tacit knowledge about different groups. Social psychology journals are weighed down with evidence showing that we understand people from our own group in a much deeper way than we do those from an "out-group." Such "out-group homogeneity" means that we see people from other groups as being much more similar to each other than they actually are. But for people from our in-group, we pay much more attention to individuality. This means that we need to work much harder to understand people from different groups.

Keeping boundaries between different aspects of our life is important. And yet when it comes to social boundaries there exists a distinct paradox. On the one hand, removing barriers between people can enrich society and our individual quality of life, but on the other, maintaining clear boundaries between different aspects of our own lives and experiences can play an important role in increasing our mental diversity. This is because it exposes us to different ways of doing things. As we know, being able to think and act in a variety of ways is essential for switch craft and helps us to become more agile. If there is a lot of overlap and similarity between different elements of our lives, this will lure us toward mental rigidity.

This has been shown in a series of studies that tell us that the greater number of separate life roles someone has, the more they are likely to thrive and be protected from depression. It's the demarcation between the roles that's particularly important. Take a woman who is a physician, married, and a mother as an example. She has three major life roles: doctor, partner, parent. She will have a range of skills and strategies from each of these roles to draw upon, and this diversity gives her a greater capacity to cope with setbacks compared to someone with just one major life role. However, let's imagine that her husband is also a doctor and works in the same medical practice. The degree of connectedness between two of her life roles has now been increased and this reduces their distinctiveness.

In her brain, the biases and contingencies that develop in one life role are not challenged to the same extent by the other. Let's illustrate this by a trivial example: imagine that both she and her husband think that a relaxing coffee break should be quiet and reflective, whereas another work colleague might prefer chatting and listening to music. At home they might always have relaxing quiet breaks, whereas at work they might be exposed to different ways to refresh. But when they are together at work, they may tend to avoid the chatty colleague and take quiet breaks together. The distinction between home and work in terms of coffee breaks now is not as different as it would be if coffee breaks were with different people in both places. Episodes like this are minor, but over time they build up and add to the mind's database over many weeks and months. We are effectively training the brain to act and think in the same way all the time and so habits get more and more deeply embedded. By doing things differently, we disrupt this habit that can help us become more alert to our surroundings—and this also disrupts the development of biases that can ultimately blind us to our environment.

Further studies have also shown that maintaining boundaries between our life roles can act as a protective buffer against stress, as an upset in one area is less likely to bleed into another. If our doctor has an argument with her husband in the morning, this will likely flow into her work life to a much greater extent than if he worked elsewhere. The more roles and experiences our doctor can maintain, the better. So, if she is also an accomplished amateur actress who rehearses with a local theater group every week and plays in a sports team on the weekend, then she will develop a richer degree of mental diversity, and an argument with her husband is then likely to have a smaller impact.

This principle has significant implications for our previous observations on resilience that we noted in Chapter 4. As we saw then, in my own research teenagers who reported the highest degrees of well-being have biases across attention, memory, and interpretation of ambiguity that were relatively disconnected from each other. While many of their biases were negative, the lack of a close *connection* between them was what really counted, because one negative bias did not necessarily activate another. The lesson is that when our negative biases remain relatively disconnected from each other, our intuitive intelligence will be richer and less biased, providing the foundations for a much greater agility of thought processes. So, while I am speculating here, ensuring that you have a range of diverse and relatively disconnected life roles and activities may help to maintain *disconnections* among your cognitive biases, further improving your intuitive powers and, of course, your switch craft.

How can we improve our intuition?

Intuition is an important pillar of switch craft because it not only tunes us in to ourselves—thus helping our self-awareness, which is another pillar of switch craft—but it also alerts us to the fine details of our environment. As we have seen throughout this chapter, there are many ways to bolster our context sensitivity. Becoming more situationally aware is a two-step process. Think of it like a complex business situation.

- First, step back mentally—as we described for decentering earlier in the book—and observe the bigger picture so that you can establish an "impressionistic" baseline.
- Then, decide what aspect of your environment you want to concentrate on, and focus in super-fine detail on that before moving on to another.

To help us tune in to our environment, the Austrian polymath Rudolf Steiner suggested a simple exercise of putting a coin on the upper left-hand side of your desk one morning, and then each morning thereafter simply moving the coin to one of the other corners of the desk. Simple though it seems, this exercise gradually leads people to develop an awareness of their immediate surroundings.

You can then begin to think about your wider context and focus in on the most relevant aspects of your situation once you have an overall sense of what's going on and what's important. This two-stage process prepares the way for greater agility.

For example, if you start a new job you could begin by focusing on how people react to the boss. Are they in awe of the boss? Do they challenge decisions? Do they do what they are asked but then complain to their colleagues when the boss is out of earshot? All these things will give you an insight into some important and often subtle aspects of the culture you have entered. Then, for another period, focus on something else; how do people interact with each other, for instance? You might make the effort to talk in more depth to one person; find out what their interests are outside work and what they are generally like. Gradually you will learn to focus better and switch efficiently from one aspect to another. And overall, you will gain a deeper understanding of your workplace and your new colleagues without overloading your brain.

Enhance your intuition

It's important to remember that intuition cannot be taught. It only emerges over time as our experience of a variety of life situations increases. We all have intuitions because, as we saw in the last chapter, this is how the brain works. It taps into its database of our past experiences, combines this with internal signals coming from our body along with signals coming from this environment, and guides us to think, feel, and act in the most appropriate way.

This is why intuition is such a crucial pillar of switch craft—these subtle guiding hints can make all the difference between getting it badly wrong and getting it right. So, while intuition cannot be taught directly, there are things we can do to help us tune in to our intuitive intelligence.

- **Quiet your mind . . . and listen:** We are surrounded by noise, whether it is constant alerts and buzzers, or the continuous chatter that goes on in our heads. All of this noise can drown out our inner intuitive voice. Your intuition cannot talk to you if you are not listening. So, make sure that you find time for solitude. By taking time to relax your body and quiet your mind you will be creating an opportunity to become more aware of your intuition. I know it's difficult in our modern, busy world, but finding even just an hour a few times a week where you can just relax, perhaps get out in nature, and learn to listen to your body is essential. A long walk, yoga, or meditation are all good ways to do this.
- **Let go of bad feelings:** We know that negative emotions are there for a reason. They alert us to problems and narrow our mind to focus on those things that are causing upset in our lives. But this works against our intuition, which requires an open mind that listens to everything. So,

it's important to learn to dial down the negative feelings—at least some of the time—to give your intuitive voice a chance to be heard. Indeed, some studies have shown that people are much better at making intuitive judgments when they are in a good mood compared to when they are in a bad mood. So use some of the tips from Chapter 10 to boost your positivity—and then listen to your intuition.

- **Look after your body:** We have seen throughout this book that our body is central to so many aspects of switch craft. We are just beginning to learn that the brain is in the service of the body, so our internal bodily signals are vital to listen to. If our body is not functioning well, it will not be able to alert us to more subtle aspects of our environment, as it will be overly focused on the basics. So, make sure that you are well nourished, get plenty of sleep, eat well, and exercise regularly. All these activities are important for many things, including allowing your intuition to find a voice.

- **Dial down your executive functions:** This is a surprising piece of advice because our executive functions, as we saw back in Chapter 7, are vital to enhance our agility. When we are analyzing situations and making rational decisions it is vital that our mental resources—inhibitory control, working memory, cognitive flexibility—are in tip-top shape. While that is certainly true, it is also true that these processes can actually work against our intuition. This sounds odd, but the logic of it is sound. When we are tired, our executive functions do not work as well and so we are more prone to distractions and less good at remembering the connections between different aspects of a situation. However, this is precisely where intuition comes into its own. When we are tired and distracted, we can be more open to new ideas and can make more creative links between things. So, at off-peak times when we are feeling tired, our intuitions may ironically have more chance of being heard.

Chapter Summary

- Context sensitivity and situational awareness come from the accumulation of knowledge as we gain more experience within a specific domain of life. This is essential to provide fuel for our intuition and, of course, is vital for switch craft.
- Broadening the diversity of our experience fine-tunes our context sensitivity and situational awareness, and reduces our cognitive biases. All of these make us more aware of our surroundings and thus better placed to make the most appropriate decisions. This is why "intuition," of which context sensitivity is a part, is such an important pillar of switch craft.
- We are all intuitive—we just need to learn to listen more carefully.
- While intuition cannot be taught—it comes with experience—there are things we can do to help us listen to our inner intuitive voice. These mainly relate to quietening the chatter in our head as well as around us, perhaps by finding times for solitude.

CONCLUSION: SOME KEY PRINCIPLES OF SWITCH CRAFT

Every spring, my husband Kevin runs a cross-country race in a remote part of the Herefordshire countryside near the Welsh border. The race is unremarkable except for the fact that it is lung-bustingly tough and attracts a fair few members of the elite Special Air Service (SAS) regiment whose camp is relatively nearby. Afterward, competitors, friends, and family gather at a pub for drinks and a post-race barbecue and, later, some live music from a local band.

A couple of years ago, just after I'd started writing *Switch Craft*, I was sitting in the garden of this country pub waiting for Kevin to appear from the "showers"—a hosepipe in a field spraying out a powerful jet of ice-cold water. It was a beautiful sunny day, and the cider was flowing freely. Next to me sat a member of the SAS who'd braved the hosepipe earlier (in other words, he'd run faster than Kevin). I knew what he did for a living—we'd been introduced earlier—but *he* wasn't familiar with what *I* did. When I told him about the book, he nodded.

I explained my golfing analogy. Sometimes you need a driver to cover the yards on the fairway. Sometimes you need a sand wedge to get you out of a bunker. Sometimes you need a putter on the green. But it's no good having the clubs without knowing when to use them. A golfer who has a full set of clubs but doesn't know which to select for the right shot would be worse off than a golfer who has just the one club in his bag.

"You've pretty much summed up Special Forces soldiering," he told me. "In the regiment, you've got lads with a bewildering array of skills you'd be hard pressed to find anywhere else in the world . . . apart from prison, maybe! See that guy over there . . ." he pointed to a giant, ginger-haired man covered in tattoos over by the food stand. "He can break into any car you put in front of him. See that guy there?" I looked at a small, dark, wiry man with a beard and a buzzcut. "He can forge any document, any signature, you want. We all work together side by side in the regiment like golf clubs in a bag."

The idea of switch craft rang a lot of bells for him. He understood that, firstly, you need to have a variety of skills. Secondly, you need to be able to adapt those skills to different situations. But thirdly, and probably most important of all, you need to have the insight to know which situation calls for which skill. And that only comes from experience.

Grit and agility

One of the unexpected paths that my career has taken me on is working with a group of elite athletes—some of these are middle-distance runners hopeful of Olympic glory. I work with Kevin, who helps the athletes to maximize mental performance when the pressure is on. When Kevin is out on the track with the athletes, I often chat with their strength-and-conditioning coach, Dan, in the hope of picking up a few tips on boosting my own fitness. "The absolute key to fitness," Dan tells me, "is to be persistent so that you build up your strength and endurance. But," he adds, "it's also important to work on your flexibility. This is the part that many people forget."

When thinking about the mental training plans that Kevin and I prepare for the athletes, I realized that we use precisely the same principles. The grittiness part of the story goes without saying. Whether you are an elite athlete or a complete beginner aiming to get from couch to 5km in eight weeks, a hefty dose of cold-blooded persistence is required. On cold, rainy nights, or first thing in the morning, it takes real effort to get out of the house to do your scheduled exercise. Persistence is essential to achieve your goal. It's no surprise, then, that there's lots written on grittiness—from scientific articles to tips in magazines to bestselling books. The importance of persistence is beyond doubt.

The significance of being agile, on the other hand, is often overlooked. But agility is arguably even more important. If an athlete picks up a minor injury, for instance, it's vital that they modify their training quickly to ensure it does not get any worse. They might do a couple of workouts on the exercise bike, for example, rather than running on the track. If they stick with their usual schedule, they run the risk of serious injury and putting themselves out of the sport for weeks or even months.

If we think about it for a moment, it becomes obvious that grit without flexibility can lead us down a path that goes nowhere. Unable to learn from our mistakes, never improving because we do not take feedback on board, we just push on regardless. Flexibility without grittiness, on the other hand, often leads to great energy in starting things and coming up with new ideas and new ways of thinking. But the problem is that a person like this rarely sticks with it and quickly gains a reputation for starting lots of different things, not being focused enough to see any of them through. Some people are so flexible in their thought processes that they are distracted by everything.

Finding the perfect balance between grit and agility is key to success in life. Switch craft can help us to do this. It's essential then to work on the four overarching pillars of the craft:

- Stay *agile* and adapt to our changing times.
- Develop your *self-awareness*.
- Work on enhancing your *emotional awareness* and ability to regulate your emotions.
- Learn to listen to your *intuition*, as this can guide you through the complexity of life and help you become more aware of your surroundings.

Agility is the main pillar of switch craft, but an agility that is supercharged by three other pillars: self-awareness, emotional awareness, and intuition. All together will ensure that you can navigate any challenge, even the most difficult ones. Succeeding in life, then, is often about finding the right balance between mental strength on the one hand and mental agility on the other, depending on the nature of the situation. The decision to stick or to switch is often crucial. And it's switch craft that will help you to make a more informed decision and will tip the scales to help you get it right more often than wrong.

Switch craft is a continuous and lifelong process. As any expert will tell you in any specialty, you never stop learning—so why should life be any different? There is no one-size-fits-all solution to any of life's problems. Many of your tried and tested tactics will work in the future, but there are times you will be faced with something completely new, and you will have to develop innovative ways to cope "on the fly." Our experiences during the coronavirus pandemic taught us this in a dramatic way. The essence of switch craft lies in developing a capacity to select the right strategy for the right moment: the right golf club for the right shot. This is often a two-step process:

- Deciding whether to stick or whether you need to switch and try something different.
- If switching, choosing the right solution for the problem you are facing.

How do we know when to stick and when to switch?

The approach we take to any situation is usually determined by the degree of *uncertainty* of that situation. If the situation is highly certain and things are going well then persisting with what we are doing is probably the best option. Why try to fix something that's not broken? As a situation becomes more uncertain, however, then we must be open to change and be agile about our approach.

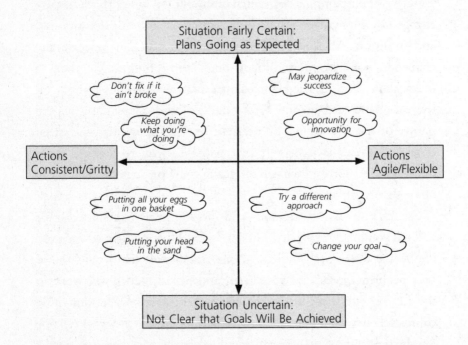

Switch craft is all about finding a way to match your approach to the nature of the situation. To do this, of course, you must have the capacity to be as gritty or as flexible as needed. And, as the SAS soldier I spoke to reminds us, to achieve this capacity you need a range of life experiences to draw upon to help you deal with a wide range of scenarios.

Your life experience gives you a rich degree of mental diversity. There is no getting away from the fact that exposure to a variety of situations and real-life experience is invaluable. There is no substitute.

This mental diversity is what provides you with a variety of options to use in volatile situations.

We have explored each of the four pillars of switch craft in detail throughout the book and you can dip into these regularly as you face different challenges. I have tried to give as many tips and suggestions as possible to help you navigate a complex and ever-changing world. Try out some of the exercises, record things in your journal if you find this helpful, and practice, practice, practice. Just like an elite athlete, practicing your craft and gaining experience in context is the most effective way to hone your expertise and maximize your chances of success.

Switch craft principles to live by

1. **Be open and curious:** Approach the world with an open and curious mind and try to avoid becoming trapped in a rigid mindset that there is only one way to do something.

2. **Become comfortable with uncertainty:** Learn to accept that the only certainty is that things change. If you avoid uncertain situations and are not willing to change, then you will gradually drift into a rigid way of thinking, feeling, and acting. You would not be alone. Many of us do become set in our ways, which is fine when life is stable and certain, but this cannot possibly last, and when things are more volatile we can rapidly be left behind with this approach.

3. **Nurture an agile way of living:** This is the first, and probably the most important, pillar of switch craft. You can become more agile by remembering the ABCD of agility: **A**dapt to changing demands; **B**alance competing desires and goals; **C**hange or challenge your perspective; and finally **D**evelop your mental competence, so that you will be able to "dance agilely in the moment," as an executive coaching friend of mine calls it.

4. **Nurture self-awareness:** This second pillar of switch craft asks you to understand your core values and be honest about your capacities. Constantly reflect on and review whether what you are doing is consistent with your authentic self. Finding a good match between who you are and your way of living is one of the keys to thriving.

5. **Accept and embrace your emotional life:** The third pillar—understanding and embracing your emotions—also fuels your agility. Some emotions don't feel great, but they are providing you with important information about the world and how you are doing. Learn to listen to what your emotions are telling you, while at the same time learning to regulate your emotions when they become overly intense. Once again, remember that *flexibility* is key; different circumstances demand different ways to regulate your emotions.

6. **Nurture your intuition:** As well as learning to listen to the internal, and often subtle, signals coming from inside your body, learn to look outside and develop a "situational awareness" of your surroundings. This fourth pillar of switch craft will allow you to make better-informed decisions and will also enhance your agility.

7. **Learn to decenter:** We have a unique ability to step back and see the bigger picture. Once we broaden our perspective, we can remind ourselves that our thoughts are not necessarily true—they are like trains passing through a station. They pass through our head, and we don't always need to engage with them or believe them. This ability to step out of the moment and see things in the bigger picture can be enormously helpful, especially in a crisis.

8. **Learn some breathing exercises and grounding techniques:** Learn a couple of simple and effective ways to relax your body. Having a basic sense of safety in your body can really help to calm your mind. We know from a wealth of research that this

is because the internal bodily signals our brain receives are constantly being interpreted and analyzed, so if your body is tense and sending messages of distress your mind will be on constant alert. Simple breathing exercises can have a surprisingly helpful effect. Try them every day—don't wait for a crisis.

9. **Create an album of happy memories:** Write down some of the happiest moments you can remember. This powerful method will provide you with happy memories to call upon in times of trouble and stress. We know from lots of research that those with a low mood find it very difficult to turn their mind to more positive thoughts. So, having an easy way to switch away from the negative to spending some time dwelling on more positive aspects of what's going on is a very good mental health strategy.

10. **Become "stress adapted":** We have seen throughout the book that experience is key. Exposing yourself to as wide a range of experiences as possible is the natural way to develop your switch craft skills. We know from research that exposure to adversity can develop our mental and social skills to deal with stress. So, rather than trying to avoid every adversity and uncertain situation, jump right in—that's the only way to learn the skills you will need to get you through future difficulties.

11. **Nurture a "wise mind":** In Dialectical behavior therapy (DBT), as we discussed in Chapter 11, there are three states of mind or ways of being. The emotional mind, where you assess a situation emotionally and intuitively; the rational mind, where you understand a situation through analyzing the objective facts; and the wise mind, which brings emotional and rational minds together. Practice asking yourself, "What would my wise mind do?" This is especially useful to do when times are calm because it will make it easier to implement when a crisis does occur.

12. **Enjoy the journey:** Remember that life can be fun. Of course, we all must deal with sadness and loss and disappointment,

these are all part of a normal life. But there are also lots of wonders out there to be enjoyed. So, embrace it—slow down, pause, and be present. Sprinkle a little awe and sparkle into your life. Look out for things to be amazed by; enjoying a great piece of art or music, noticing the beauty of nature, or observing the vastness of the night sky, are all great ways of putting everything into perspective.

Switch craft helps us to cope in an uncertain world

Switch craft is a supercharged agility that allows us to thrive in a precarious world where many moving pieces come at us from unexpected directions. An unwavering focus on grit means we risk missing the intrinsic richness, creativity, and wildness of life. There are many examples of how people who are generalists—those who do not necessarily persist with one thing, whether it be an occupation or a hobby—are often more successful in, and happy with, their lives. The important thing is finding the right match. It is not quitting or switching for the sake of it that's important, it is finding the best match for your talents and interests.

A friend of mine, Jonathan, found this out the hard way. He was passionate about science in school and dreamed of studying chemistry at university. His family, however, had come from a long line of lawyers and he was under immense pressure to also go into law. He bowed to family pressure and struggled through a law degree, hating every minute. Years working in a solicitor's office followed, including training to become a barrister. Jonathan was not a natural storyteller and could only watch in wonder as his peers skillfully manipulated juries with the brilliance of their narrative.

Ironically, his life changed when he was part of a defense team for a chemist who was claiming ownership of a patent that his university employer said rightfully belonged to them. Hours of questioning and discussing the details of the invention with the client reignited Jonathan's love of chemistry. A few months after that case he made the extraordinary decision to quit the law and retrain in science. He has never looked back and only wishes that he had made the leap much earlier. "Thank goodness I was brave enough to quit," he told me. "Otherwise it would have been a long slow road to unhappiness."

Quitting was the right decision for Jonathan. And it often is. Downsizing his dental business was the best solution for Paddi Lund. My own decision to quit accounting all those years ago was, almost certainly, right for me. When I lay sobbing in my bedroom at the age of seventeen, I thought my future was doomed. Instead, the journey was just beginning!

None of us will ever know for sure whether we've taken the right path, or made the right decision, of course, because everyday problems do not have obvious correct answers. But what we do know is that the inherent uncertainty of life requires switch craft to succeed.

I hope this book will help to set you on a more resilient pathway. By learning to understand and accept yourself, appreciate your surroundings, and recognize the power of being agile and keeping an open mind, my hope is that you will be ready to embrace the rest of your life not as a chore, but as an adventure.

APPENDIX 1: SOLUTION TO NINE-DOT PROBLEM

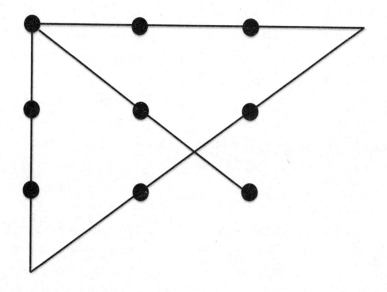

APPENDIX 2: HOW TO SCORE THE COMPONENTS OF A PERSONAL NARRATIVE

These ratings have been adapted from a scientific paper by Kate C. McLean and her colleagues, "The empirical structure of narrative identity: The initial Big Three," *Journal of Personality and Social Psychology*, 2020, vol. 119, pp. 920–944. Further background is provided in a research primer by Jonathan Adler and colleagues published in the journal *Social Psychological and Personality Science*, 2017, vol. 8, pp. 519–27.

Emotional Quality:

Agency (A): This relates to the degree of autonomy you have in your narrative. Did you have much control over the situation? Give yourself a score of 0 if you were at the mercy of the situation and powerless to act, and then 1, 2, 3, or 4, with 4 referring to total control over the situation and a clear demonstration that you were able to initiate change on your own and affect your life.

Communion (C): This scores 0 if you were completely disconnected, isolated, or rejected, and then 1 to 4, with 4 being evidence of high connections with others and a rich way of describing these connections.

Emotional tone (E): This relates to the emotional feel of the overall narrative and it is rated on a 1–5 scale with 1 being very pessimistic or negative, 5 being very optimistic or positive, and 3 being emotionally neutral.

Redemption (R) and Contamination (Ct): Redemption is defined as a narrative that begins in a negative state (e.g. loss or illness) and ends in a positive state, while contamination starts in a positive or neutral state but ends in a negative state. No evidence of change from beginning to end should be given a 0. Evidence of a small change from negative to positive should be given a 1, while a larger shift to a positive state is given a 2. Evidence for contamination—moving from a neutral or positive state to a negative one—is given minus 1.

The calculation of Emotional Quality is $(A + C + E + R) - Ct$. In other words, simply add up your scores on three of the components (A, C, and E) and then add a score for R if this was a positive score or take away 1 if your narrative had a contamination theme (went from good to bad). SCORE =

Meaning-Making:

Exploratory Processing (EP): Does your story outline an open analysis and exploration of the meaning of past events in order to understand the impact of the event and its potential to change your sense of self? This might be in the form of reflecting on how you felt during the event, whether you discussed the nature of the event with others, and whether you have outlined explicitly how you might have been changed by the event. This is rated on a 5-point scale from 0–4 with 0 being no evidence for any exploration of the event, 1 being minimally exploratory, and 4 being highly exploratory, with exploration being the primary theme of the narrative.

Meaning-Making (MM): This is the degree to which you have increased your self-knowledge and gained some insights by reflecting on the past experience. A rating of 0 reflects no explanation of the meaning of the event; 1 is given if a specific lesson was learned; a score of 2 is given for narratives that contained vague meaning (e.g. some growth or change in the self is reported but without specifics); and 3 is given if there is evidence that you gained specific insights from the event that you then applied to wider areas of your life.

Change Connections (CC): This component reflects the degree to which the event or situation narrated led to a change in some aspect of your self-understanding. A score of 0 is given if there is no evidence that the experience led to any change in self-understanding; a 1 is given for evidence that the event caused a change in self-understanding; and a further point (giving 2 in total) is given if the event brought to light a previously unknown aspect of self.

Growth (G): This is the degree to which a sense of positive personal growth was described as an outcome of the event narrated. A rating of 0 is given if there is no evidence at all for any personal growth; a rating of 1 is given if the narrative displayed some suggestion of growth that was minimally elaborated; a rating of 2 is given if positive growth was clearly present and described as important or transformative; and a rating of 3 is given if growth is a highly detailed theme of the narrative and the growth described was clearly transformative.

The calculation of Meaning-Making is (EP + MM + CC + G). In other words, simply add up your scores on the four components. SCORE =

Complexity:
Key components of complexity are factual details along with the coherence of the story.

Facts (F): These are the factual details of the situation—such as where it happened, when it happened, who did what, and so on. A rating of 0 is given for a complete absence of factual details, with a rating of 1–3 being given depending on the degree of factual details that transport the reader to the event. A score of 3 would often include rich details about motivations, intentions, and internal states that offer some insight into how you were feeling or thinking at the time.

Coherence (Co): This is the degree to which the narrative makes sense in terms of timing and context. Are there clear details of the situation (context) and a lucid sense of the timescale of what happened? A complete absence of coherence to the story is given a 0, with a 1 being given for clear details of the time and place of the event (context) and a further point (giving 2 in total) being given for a clear outline of the temporal organization of the event.

The calculation of Complexity is F + Co. In other words, simply add up your scores on the two components. SCORE =

Emotional Quality: Possible scores go from 0 to 15
Meaning-Making: Possible scores go from 0 to 12
Complexity: Possible scores go from 0 to 5

NOTES

Introduction

"We often lose this fluidity . . ." (p. 4): There is evidence that being able to remain mentally agile is linked with higher well-being as we grow older. See Julie Blaskewicz Boron and colleagues, "Longitudinal change in cognitive flexibility: Impact of age, hypertension, and APOE4," *Innovation in Aging*, 2018, vol. 2, p. 249.

"Our brains have evolved to operate as 'prediction machines.'" (p. 5): In the twentieth century neuroscientists thought that the brain learned by extracting information from sensory signals coming in from the outside world. This changed completely in the twenty-first century when neuroscience shifted to seeing the brain as an "inference device" that actively constructed explanations and predictions of what's going on in the "outside world." A highly accessible explanation of how this works can be found in Lesson 4 of a fascinating book by neuroscientist and psychologist Lisa Feldman Barrett, *Seven and a Half Lessons about the Brain*, Picador, 2020. If you would like a more academic read, a good scientific overview of this new perspective is available in a paper by Karl Friston: "Does predictive coding have a future?" *Nature Neuroscience*, 2018, vol. 21, pp. 1019–21.

"A surprising finding in the science of emotion . . ." (p. 5): For an overview of "feeling tones" from a Buddhist perspective see "Mindfulness theory: feeling tones (vedanãs) as a useful framework for research" by Martine Batchelor, *Current Opinion in Psychology*, 2019, vol. 28, pp. 20–22. More modern work in affective science is beautifully explained by Lisa Feldman

Barrett in her book *How Emotions Are Made: The Secret Life of the Brain*, Macmillan, 2017.

"An inflexible mind leads to anxiety and depression." (p. 6): Psychologists Todd Kashdan and Jonathan Rottenberg have accrued much evidence that inflexibility of mental processes is at the heart of many psychological problems. Their review of the literature in 2010 did much to kick-start this field of research. See Todd Kashdan and Jonathan Rottenberg, "Psychological flexibility as a fundamental aspect of health," *Clinical Psychology Review*, 2010, vol. 30, pp. 865–78.

"I have always been fascinated by how our attention is captured by negative information." (p. 6): These studies are described in several scientific papers, including: Elaine Fox, "Attentional bias in anxiety: Selective or not?" *Behavior Research and Therapy*, 1993, vol. 31, pp. 487–93 , and Elaine Fox, "Allocation of visual attention and anxiety," *Cognition and Emotion*, 1993, vol. 7, pp. 207–15.

"I was never convinced that this 'high alert' theory was the whole story." (p. 7): Some of these studies are described in the scientific paper by Elaine Fox, Riccardo Russo, Robert Bowles, and Kevin Dutton: "Do threatening stimuli draw or hold visual attention in subclinical anxiety?" *Journal of Experimental Psychology*, 2001, vol. 130, pp. 681–700.

"These recommendations are backed up by solid science . . ." (p. 8): There are several compelling books by leading experts describing the extensive evidence base for mindfulness (Mark Williams and Danny Penman, *Mindfulness: A Practical Guide to Finding Peace in a Frantic World*, Piatkus Books, 2011; Ruby Wax, *A Mindfulness Guide for the Frazzled*, Penguin Life, 2016), grittiness (Angela Duckworth, *Grit: The Power of Passion and Perseverance*, Vermillion, 2016), growth mindset (Carol Dweck, *Mindset: The New Psychology of Success*, Ballantine Books, 2007), and positivity (Barbara Fredrickson, *Positivity: Groundbreaking Research to Release Your Inner Optimist and Thrive*, Oneworld Publications, 2011).

"There is much evidence that we need a range of approaches . . ." (p. 8): The American journalist David Epstein has written a compelling and best-selling book demonstrating the importance of sampling widely from life to maximize your opportunities in a complex world: David Epstein, *Range: How Generalists Triumph in a Specialized World*, Riverhead Books, 2019.

Chapter 1: Accepting Change and Adapting to It

"Colonel Pete Mahoney commands . . ." (p. 15): I have heard Pete Mahoney speak at several meetings on resilience, and he never fails to enthrall with his stories of remarkable resilience from the front line.

"Instead of creating a false divide." (p. 18): For many years I was part of the academic board of the global executive coaching firm, MindGym, and our mantra when advising businesses was that change is business as usual. You can read more about the MindGym approach in a book by its two founders: Sebastian Bailey and Octavius Black, *MindGym: Achieve More by Thinking Differently*, HarperOne, 2016.

"Changes and transitions" (p. 20): Many books have been written about how to adapt well to change. Two I have found particularly useful are William Bridge's *Transitions: Making Sense of Life's Changes*, Da Capo Lifelong Books, dated edition, 2020, and Julia Samuel's *This Too Shall Pass*, Penguin Life, 2020.

"The German psychoanalyst Fritz Perls . . ." (p. 22): Fritz Perls, *The Gestalt Approach and Eye Witness to Therapy*, Science and Behavior Books, 1989.

"Extensive research on quitting smoking . . ." (p. 26): James Prochaska and Carlo DiClemente, "Stages and processes of self-change of smoking: toward an integrated model of change," *Journal of Consulting and Clinical Psychology*, 1983, vol. 51, pp. 390–95.

"Often it is about small steps . . ." (p. 29): The website www.jamesclear.com is packed full of great tips and suggestions as to how you can build and maintain good habits step by step, as is James Clear's bestselling book *Atomic Habits*, Random House Business Books, 2018.

"'Be like water,' Lee tells us." (p. 30): You can find several interviews with Bruce Lee on YouTube. There is also a wonderful overview of his lesser-known views in an article by Maria Popova, "Bruce Lee's never before seen writings on Willpower, Emotion, Reason, Memory, Imagination, and Confidence," on her uplifting *The Marginalian* website (www.themarginalian.org).

Chapter 2: Managing Uncertainty and Worry

"Rituals can bring some structure and order . . ." (p. 33): Martin Lang provides a nice overview of the function of rituals from scientific and religious perspectives in "The evolutionary paths to collective rituals: An interdisciplinary perspective on the origins and functions of the basic social act," *Archive for the Psychology of Religion*, 2019, vol. 41, pp. 224–52. Sasha Sagan, in her heartwarming book *For Small Creatures Such as We*, Penguin Random House, 2019, has also argued that rituals are essential to help us to find meaning in our fragile existence.

"The following questions will give you an idea . . ." (p. 36): The questions presented here are a modified version of a standardized "intolerance of uncertainty" questionnaire developed by R. Nicholas Carleton and his colleagues: R. Nicholas Carleton, Peter J. Norton, and Gordon J. G. Asmundson, "Fearing the unknown: A short version of the Intolerance of Uncertainty Scale," *Journal of Anxiety Disorders*, 2007, vol. 21, pp. 105–17.

"Psychologists have been researching . . ." (p. 37): R. Nicholas Carleton et al., "Increasing intolerance of uncertainty over time: the potential influence of increasing connectivity," *Cognitive Behavior Therapy*, 2019, vol. 48, pp. 121–36.

"This is why worrying is a common by-product . . ." (p. 38): Michel Dugas and his colleagues have produced a large body of work demonstrating that intolerance of uncertainty leads to chronic worry and anxiety. An early version of their model can be found in Michel Dugas, Mark Freeston, and Robert Ladouceur, "Intolerance of uncertainty and problem orientation in worry," *Cognitive Therapy and Research*, 1997, vol. 21, pp. 593–606.

"Worry that leads to action can be productive . . ." (p. 39): You can find a highly readable overview of the difference between productive and unproductive worry in Robert Leahy, *The Worry Cure: Seven Steps to Stop Worry Stopping You*, Harmony Books, 2006.

"Numerous research studies tell us that uncertainty pushes us toward the familiar . . ." (p. 41): A large body of research led by Polish psychologist Arie Kruglanski, now based in the USA, has shown that feeling uncertain leads to what he calls a strong "need for closure." This has multiple implications including pushing us to make quick decisions, to choose the familiar over the unfamiliar, to favor our own group over others, and to close our mind to alternative possibilities. An early academic overview of this idea can be found in Arie Kruglanski and Donna Webster, "Motivated closing of the mind: seizing and freezing," *Psychological Review*, 1996, vol. 103, pp. 263–83.

"In one intriguing set of studies . . ." (p. 41): Edward Orehek et al., "Need for closure and the social response to terrorism," *Basic and Applied Social Psychology*, 2010, vol. 32, pp. 279–90.

"They work by exposing people to small 'doses' of uncertainty . . ." (p. 44): Michel Dugas and Robert Ladouceur, "Treatment of Generalized Anxiety Disorder: Targeting Intolerance of Uncertainty in Two Types of Worry," *Behavior Modification*, 2000, vol. 24, pp. 635–57.

"Consider planning some behavioral experiments of your own . . ." (p. 44): If you are very anxious about uncertainty, you might want to consider exploring behavioral experiments further with a skilled therapist. A typical treatment package takes from fourteen to about sixteen sessions.

"One surprisingly effective way to do this." (p. 47): You can record your worries in your journal or on your phone. There is also an easy-to-use and free app, Worry Tree, that is available to help you record, manage ,and problem-solve your worries: https://www.worry-tree.com/worrytree -mobile-app.

"This is called, decentering," (p. 49): Decentering is the ability to notice everyday stressful situations from an objective, third-person perspective. This simple technique has been shown to be very powerful in reducing anxiety and depression. For a good academic overview, see a paper by Marc Bennett et al., "Decentering as a core component in the psychological treatment and prevention of youth anxiety and depression: A narrative review and insight report," *Translational Psychiatry*, 2021, vol. 11, article 288.

Chapter 3: The Flexibility of Nature

"These worms are still studied in laboratories all over the world . . ." (p. 51): *Caenorhabditis elegans* is a transparent roundworm—or nematode— about 1mm long. Sydney Brenner, along with his colleagues John Sulston and Robert Horvitz, were awarded the 2002 Nobel Prize in Physiology or Medicine for their work on *C. elegans*.

"Brenner and his team discovered many fascinating things about the brain . . ." (p. 52): The nervous system actually contains two types of cells: *neurons*, which process information, and *glia*, which provide the neurons with metabolic as well as mechanical support. When I refer to "brain cells" I mean the neurons.

"What has been discovered in much more recent research . . ." (p. 52): Yuan Wang et al., "Flexible motor sequence generation during stereotyped escape responses," *eLife*, 2020, article 56942.

"Almost all fish . . ." (p. 52): There is a startling demonstration of the remarkable ability of fish to change their sex in the first episode of the BBC nature program *Blue Planet II*, along with an interesting explanation of the biology involved: https://www.bbcearth.com/news/fish-are-the-sex-switching-masters-of-the-animal-kingdom.

"At a more fundamental level are bacteria . . ." (p. 53): Stealing genes as an adaptive strategy is highlighted in a brief article by Mitch Leslie, "Stealing genes to survive," *Science*, 2018, vol. 359, p. 979.

"By producing a defense for every possible scenario . . ." (p. 54): The immune system is remarkably adaptable and flexible. Biologist Gerald Edelman, along with his colleague Rodney Porter, received the 1972 Nobel Prize for their discoveries that each time a cell divided, tiny errors would occur, which resulted in the production of multiple slightly different proteins, and it was this variety that the immune system took advantage of to fight foreign bodies. In recent research, an Australian team have discovered that the immune system can mutate certain types of its cells (B cells) to create antibodies that are more specifically tweaked to foreign bodies when the targets for those bodies, called "antigens," are flexible rather than rigid: Deborah L. Burnett et al., "Conformational diversity facilitates antibody mutation trajectories and discrimination between foreign and self-antigens," *Proceedings of the National Academy of Science*, 2020, vol. 117, pp. 22341–350.

"This tendency of biological systems to do the same thing . . ." (p. 54): Gerald Edelman and Joseph Gally, "Degeneracy and complexity in biological systems," *Proceedings of the National Academy of Sciences*, 2001, vol. 98 (24), pp. 13763–68.

"Compared to the 302 neurons and 8,000 synapses . . ." (p. 54): We know about the number of brain cells (glia and neurons) in the human brain because of the remarkable work of Brazilian neuroscientist Suzana Herculano-Houzel. You can read about her fascinating work in a highly accessible read from the scientist herself in *The Human Advantage: A New Understanding of How Our Brains Became Remarkable*, MIT Press, 2016.

"This is referred to as the Hebbian rule . . ." (p. 55): Donald Olding Hebb published his influential book *The Organization of Behavior* in 1949. It still provides a powerful framework of understanding how the brain works and is considered to be one of the most influential books ever published in psychology and neuroscience.

"Neuroscientist Lisa Feldman Barrett describes this . . ." (p. 56): The Boston-based psychologist and neuroscientist provides an engaging and fascinating overview of how our brain works and the importance of what she calls the "body budget" in her book *Seven and a Half Lessons About the Brain*, Picador, 2020.

"Think about the tennis player Roger Federer . . ." (p. 57): Journalist Oliver Pickup wrote an excellent overview of the art of returning the serve in elite men's tennis in the *Telegraph* newspaper, which you can read here: https://www.telegraph.co.uk/tennis/wimbledon-reaction/how-to-return-a-serve.

Chapter 4: Agility and Resilience

"We now know that resilience is an ongoing and dynamic process . . ." (p. 59): An expert group of researchers published an important guide to improve resilience research in 2017. They emphasized the importance of capturing the dynamic nature of adaptation to stress in long-term studies: Raffael Kalisch et al., "The resilience framework as a strategy to combat stress-related disorders," *Nature Human Behavior*, 2017, vol. 1, pp. 784–90.

"Study after study shows that most of us . . ." (p. 60): The New York psychologist George Bonanno has shown that there are many different trajectories that people experience for several years following a major trauma and that resilience is the most common outcome: George Bonanno, "Loss, Trauma, and Human Resilience: Have We Underestimated the Human Capacity to Thrive After Extremely Aversive Events?" *American Psychologist*, 2004, vol. 59, pp. 20–28.

"Resilience has as much to do with." (p. 60): The coronavirus pandemic did not have the adverse effect on mental health that many were expecting. Instead, people the world over were highly resilient and adapted well to the changes. You can read about this research in an article for a magazine by psychologists Lara Aknin, Jamil Zaki, and Elizabeth Dunn, "The Pandemic Did Not Affect Mental Health the Way You Think," *The Atlantic*, July 4, 2021.

"In fact, recent research tells us . . ." (p. 60): Jessica Fritz et al., "A systematic review of amenable resilience factors that moderate and/or mediate the relationship between childhood adversity and mental health in young people," *Frontiers in Psychiatry*, 2018, vol. 8, article 230.

"Leading resilience researcher Michael Ungar . . ." (p. 61): I spoke with the

Canadian resilience researcher Michael Ungar following an MQ Mental Health Science meeting in London in February 2019. The various ways in which we can recover and adapt positively to adversity is nicely summarized in his book *Change Your World: The Science of Resilience and the True Path to Success*, Sutherland House, 2019.

"Several studies of resilience among refugees . . ." (p. 61): Emrah Cinkara, "The role of L+ Turkish and English learning in resilience: A case of Syrian students at Gaziantep University," *Journal of Language and Linguistic Studies*, 2017, vol. 13, pp. 190–203. A good overview of the importance of language in enhancing resilience for refugees can be found in a report from the British Council, "*Language for Resilience*": www.britishcouncil.org/language-for-resilience.

"Studies by my lab group have explored resilience . . ." (p. 63): In our studies with a large cohort of around five hundred teenagers we found a variety of protective influences: these included being male, coming from an affluent background, as well as certain types of thinking styles and levels of self-esteem. You can read the results of our research in Charlotte Booth, Annabel Songco, Sam Parsons, and Elaine Fox, "Cognitive mechanisms predicting resilient functioning in adolescence: Evidence from the CogBIAS longitudinal study," *Development and Psychopathology*, 2020, pp. 1–9.

"However, there is also evidence that some of the related traits . . ." (p. 63): Some research has suggested that the pay gap that still exists between men and women may be due to a backlash against women when they try to negotiate for higher salaries, which is often seen as gender inappropriate. Jennifer Dannals, Julian Zlatev, Nir Halevy, and Margaret Neale, "The dynamics of gender and alternatives in negotiation," *Journal of Applied Psychology*, 2021, https://doi.org/10.1037/apl0000867.

"And there is a mountain of research . . ." (p. 64): There are many studies exploring the costs and benefits of problem-focused and emotion-focused coping styles. One interesting study that shows the complexity of the issue is John Baker and Howard Berenbaum, "Emotional approach and problem-focused coping: A comparison of potentially adaptive strategies," *Cognition and Emotion*, 2007, vol. 21, pp. 95–118.

"There is a growing body of research telling us . . ." (p. 65): The Hong Kong psychologist Cecilia Cheng has led the way in demonstrating the importance of coping *flexibility* in building well-being and resilience: Cecilia Cheng and Chor-Iam Chau, "When to approach and when to avoid?:

Functional flexibility is the key," *Psychological Inquiry*, 2019, vol. 30, pp. 125–29, and Cecilia Cheng, Hi-Po Bobo Lau, and Man-Pui Sally Chan, "Coping flexibility and psychological adjustment to stressful life changes: A meta-analytic review," *Psychological Bulletin*, 2014, vol. 140, pp. 1582–1607.

"This is backed up by our own studies . . ." (p. 65): Charlotte Booth, Annabel Songco, Sam Parsons, and Elaine Fox, "Cognitive mechanisms predicting resilient functioning in adolescence: Evidence from the CogBIAS longitudinal study," *Development and Psychopathology*, 2020, pp. 1–9, and Sam Parsons, Anne-Wil Kruijt, and Elaine Fox, "A cognitive model of psychological resilience," *Journal of Experimental Psychopathology*, 2016, vol. 7, pp. 296–310.

"But the deficit model misses interesting information . . ." (p. 65): The work of psychologist and anthropologist Bruce Ellis and his team at the University of Utah has suggested that looking for young people's strengths is often more informative than looking for their deficits: Bruce Ellis et al., "Beyond risk and protective factors: An adaptation-based approach to resilience," *Perspectives on Psychological Science*, 2017, vol. 12, pp. 561–87.

"In the early 1980s Jason Everman blew up a toilet at his junior high school . . ." (p. 67): Clay Tarver, "The Rock 'n' Roll Casualty Who Became a War Hero," *New York Times*, July 2, 2013.

Chapter 5: The Benefits of Mental Agility

"Paddi realized that he had to do something radical." (p. 73): I heard Paddi speak about his radical move at a business conference several years back. You can read about how he built his happiness-centered dental practice in his digital e-book *Building the Happiness-Centered Business*, which can be downloaded from https://www.paddilund.com.

"As the American psychologist Abraham Maslow warned . . ." (p. 74): Abraham Maslow, *The Psychology of Science*, Harper and Row, 1966.

"Recent and cutting-edge science has shown. . . ." (p. 78): Psychologists Todd Kashdan and Jonathan Rottenberg have argued that the ability to be psychologically flexible is associated with health and well-being. You can read their extensive review of psychology studies in Todd Kashdan and Jonathan Rottenberg, "Psychological flexibility as a fundamental aspect of health," *Clinical Psychology Review*, 2010, vol. 30, pp. 865–78. More recent studies have shown that flexibility is important for boosting mental health during a time of crisis. A good example is David Dawson and Nima Golijani-Moghaddam, "COVID-19: Psychological flexibility, coping, mental health, and well-being in the UK during the pandemic," *Journal of Contextual Behavioral Science*, 2020, vol. 17, pp. 126–34.

"Let me explain by looking at one of my own studies . . ." (p. 79): Sam Parsons, Annabel Songco, Charlotte Booth, and Elaine Fox, "Emotional information-processing correlates of positive mental health in adolescence: A network analysis," *Cognition and Emotion*, 2021, vol. 35, pp. 956–969.

"One way to foster these more flexible connections . . ." (p. 80): A good overview of how interpretation bias can influence well-being and how it can be modified is available in a chapter by Courtney Beard and Andrew Peckham, "Interpretation bias modification," Chapter 20 of *Clinical Handbook of Fear and Anxiety: Maintenance Processes and Treatment Mechanisms*, edited by Jonathan Abramowitz and Shannon Blakey, American Psychological Association, 2020.

"In the late 1990s and early 2000s, LEGO was in real trouble." (p. 83): You can learn all you want to know about how LEGO reinvented itself and overturned the rules of innovation in *Brick By Brick: How LEGO Rewrote the Rules of Innovation and Conquered the Global Toy Industry* by Wharton Professor of Innovation, David Robertson, Crown Business, 2013.

"Despite its apparent simplicity, this puzzle is fiendishly difficult." (p. 85): Trina Kershaw and Stellan Ohlsson, "Multiple causes of difficulty in insight: The case of the nine-dot problem," *Journal of Experimental Psychology: Learning, Memory & Cognition*, 2004, vol. 30, pp. 3–13.

"In medieval times, contagious diseases and epidemics . . ." (p. 85): Marianna Karamanou, George Panayiotakopoulos, Gregory Tsoucalas, Antonis Kousoulis, and George Androutsos, "From miasmas to germs: A historical approach to theories of infectious disease transmission," *Le Infezioni in Medicina*, 2012, vol. 1, pp. 52–56.

"But psychologists from both Dartmouth and Princeton started to wonder . . ." (p. 88): Albert Hastorf and Hadley Cantril, "They saw a game: A case study," *Journal of Abnormal and Social Psychology*, 1954, vol. 49, pp. 129–34.

"It's why we are much more likely to notice wrongdoing in strangers . . ." (p. 88): Ursula Hess, Michel Cossette, and Shlomo Hareli, "I and my friends are good people: The perception of incivility by self, friends and strangers," *European Journal of Psychology*, 2016, vol. 12, pp. 99–114.

Chapter 6: The Nuts and Bolts of Agility
in the Brain: Cognitive Flexibility

"Vestiges of this reaction remain in our own brains . . ." (p. 91): I have made this point in my own work when describing the delay in disengaging from threat that we see in anxious people as a kind of mini brain freeze: Elaine Fox et al., "Do threatening stimuli draw or hold visual attention in subclinical anxiety?" *Journal of Experimental Psychology*, 2001, vol. 130, pp. 681–700.

"Two different internal processes in your brain . . ." (p. 92): Diana Armbruster et al., "Prefrontal cortical mechanisms underlying individual differences in cognitive flexibility and stability," *Journal of Cognitive Neuroscience*, 2012, vol. 24, pp. 2385–99.

"What brain imaging studies also show is that agile people . . ." (p. 93): Urs Braun et al., "Dynamic reconfiguration of frontal brain networks during executive cognition in humans," *Proceedings of the National Academy of Sciences*, 2015, vol. 112, pp. 11678–83.

"In our brain, cognitive flexibility . . ." (p. 94): In psychology, an experimental paradigm known as "task switching" is used to measure cognitive flexibility. A clear overview is available in an article by the British psychologist Stephen Monsell, "Task switching," *Trends in Cognitive Sciences*, 2003, vol. 7, pp. 134–40.

"My argument is that this ability . . ." (p. 94): The nature and definition of "cognitive flexibility" varies in different psychology studies. I take a more general approach and incorporate most of these definitions in this book. A good academic overview of the nature of cognitive flexibility can be found in a theoretical piece by the Romanian psychologist, Thea Ionescu, "Exploring the nature of cognitive flexibility," *New Ideas in Psychology*, 2012, vol. 30, pp. 190–200.

"Between the ages of about seven and eleven . . ." (p. 94): A good overview of how cognitive flexibility and other executive functioning skills develop across childhood can be found in a research study by the Illinois State University team of Alison Bock, Kristin Gallaway, and Alycia Hund, "Specifying links between executive functioning and theory of mind during middle childhood: Cognitive flexibility predicts social understanding," *Journal of Cognition and Development*, 2015, vol. 16, pp. 509–21.

"A good way to test this is to ask children . . ." (p. 95): This is an example of a "Multiple Classification Sorting Task": B. Inhelder and Jean Piaget,

The Early Growth of Logic in the Child: Norton Books, 1964. Variations of this task have been used in many studies to measure cognitive flexibility in children.

"Indeed, several studies tell us . . ." (p. 95): Children who are good at these type of "task switching" games are also less likely to use rigid stereotypes: Rebecca Bigler and Lynn Liben, "Cognitive mechanisms in children's gender stereotyping: Theoretical and educational implications of a cognitive-based intervention," *Child Development*, 1992, vol. 63, pp. 1351–63. If they are good at this task, children also tend to be good at reading and other essential cognitive skills. For further details see Kelly Cartwright, "Cognitive development and reading: The relation of reading-specific multiple classification skill to reading comprehension in elementary school children," *Journal of Educational Psychology*, vol. 94, pp. 56–63, and Pascale Cole, Lynne Duncan, and Agnes Blaye, "Cognitive flexibility predicts early reading skills," *Frontiers in Psychology*, 2014, vol.5, p. 565.

"While there is a mixed bag of results . . ." (p. 100): There is some evidence that video action games that emphasize rapid switching among multiple activities can lead to large improvements in cognitive flexibility: Kerwin Olfers and Guido Band, "Game-based training of flexibility and attention improves task-switch performance: Near and far transfer of cognitive training in an EEG study," *Psychological Research*, 2018, vol. 82, pp. 186–202, and Brian Glass, W. Todd Maddox, and Bradley Love, "Real-time strategy game training: emergence of a cognitive flexibility trait," *PLOS One*, August 7, 2013, vol. 8 (8), e70350.

"Another way to boost your cognitive flexibility is to travel." (p. 100): Several psychology studies support the idea that travel does broaden the mind and, more specifically, improves our cognitive flexibility. A highly accessible overview is available in an article written for the *Harvard Business Review* by American psychologist Todd Kashdan: "Mental benefits of vacationing somewhere new," https://hbr.org/2018/01/the-mental-benefits-of-vac ationing-somewhere-new.

"One research team examined the creativity of senior designers . . ." (p. 100): Frederic Godart, William Maddux, Andrew Shipilov, and Adam Galinsky, "Fashion with a foreign flair: Professional experiences abroad facilitate the creative innovations of organizations," *Academy of Management Journal*, 2015, vol. 58, pp. 195–220.

"The 'Unusual Uses Test' . . ." (p. 101): Robert Wilson, J. P. Guilford, Paul

Christensen, and Donald Lewis, "A factor-analytic study of creative thinking abilities," *Psychometrika*, 1954, vol. 19, pp. 297–311.

"However, as our stress levels increase . . ." (p. 102): Studies have shown that increased anxiety leads to greater difficulty in dragging our attention away from a very engaging task: Daniel Gustavson, Lee Altamirano, Daniel Johnson, Mark Whisman, and Akira Miyake, "Is set-shifting really impaired in trait anxiety? Only when switching away from an effortfully established task set," *Emotion*, 2017, vol. 17, pp. 88–101.

"But internally a different story emerges . . ." (p. 102): This was found in an important study conducted by the Birkbeck College, University of London, psychologists Tahereh Ansari and Nazinin Derakshan, "The neural correlates of cognitive effort in anxiety: Effects on processing efficiency," *Biological Psychology*, 2011, vol. 86, pp. 337–48.

"Instead of using the traditional version of task switching . . ." (p. 103): An affective version of the task-switching test was developed by the University of Miami–based psychologists Jessica Genet and Matthias Siemer. This task measures a person's ability to switch between the emotional aspects of a word or a picture as opposed to the nonemotional aspects: Jessica Genet and Matthias Siemer, "Flexible control in processing affective and non-affective material predicts individual differences in trait resilience," *Cognition and Emotion*, 2011, vol. 25, pp. 380–88. I designed this study along with a student, Eve Twivy, who was conducting her MSc research in my lab group, and a Dutch psychologist, Maud Grol, who was a post-doctoral researcher in my lab group at the time.

"Given my own earlier findings . . ." (p. 103): Elaine Fox, Riccardo Russo, Robert Bowles, and Kevin Dutton, "Do threatening stimuli draw or hold visual attention in subclinical anxiety?" *Journal of Experimental Psychology*, 2001, vol. 130, pp. 681–700.

"An inflexibility to switch away from negative material . . ." (p. 103): Jessica Genet, Ashley Malooly, and Matthias Siemer, "Flexibility is not always adaptive: affective flexibility and inflexibility predict rumination use in everyday life," *Cognition and Emotion*, 2013, vol. 27, pp. 685–95.

"To better understand how people coped with everyday pressures . . ." (p. 104): The Hassles and Uplifts Scale (HUS) provides an assessment of how many daily irritants (like breaking a watch strap, missing the bus) a person experiences in an average week along with uplifts (such as phoning a friend, going out to eat, finishing a task). It was developed by Anita DeLongis et al., "Relationships of daily hassles, uplifts and major life events to health status," *Health Psychology*, 1982, vol. 1, pp. 119–36.

"While there were some signs of increased rigidity . . ." (p. 104): You can read about our results in Eve Twivy, Maud Grol, and Elaine Fox, "Individual differences in affective flexibility predict future anxiety and worry," *Cognition and Emotion*, 2021, vol. 35, pp. 425–34.

"Psychological agility in this wider sense . . ." (p. 106): Psychologists Todd Kashdan and Jonathan Rottenberg have written a seminal review of the academic literature showing the benefits of psychological flexibility for health and well-being: "Psychological flexibility as a fundamental aspect of health," *Clinical Psychology Review*, 2010, vol. 30, pp. 865–78.

Chapter 7: The ABCD of Mental Agility

"Broader psychological agility consists of four dynamic processes . . ." (p. 111): These four elements were identified in an extensive review of psychology studies by Todd Kashdan and Jonathan Rottenberg: "Psychological flexibility as a fundamental aspect of health," *Clinical Psychology Review*, 2010, vol. 30, pp. 865–78.

"As the British entrepreneur . . ." (p. 113): See https://www.virgin.com /branson-family/richard-branson-blog/my-top-10-quotes-on-change.

"The science shows us that watching someone like ourselves . . ." (p. 115): The Canadian-American psychologist Albert Bandura spent a lifetime developing his theory of "self-efficacy" and found that we learn to succeed by observing others make a sustained effort to succeed. He called this "social modelling," where we take on the habits of the people we are surrounded by: Albert Bandura, *Social Learning Theory*, Prentice Hall, 1977.

"A good example is a study by a team of Spanish and British researchers . . ." (p. 115): Ana Isabel Sanz-Vergel, Alfredo Rodriguez-Munoz, and Karina Nielson, "The thin line between work and home: The spillover and cross-over of daily conflicts," *Journal of Occupational and Organizational Psychology*, 2014, vol. 88, pp. 1–18.

"Even though it might take up considerable time." (p. 117): Several studies have shown the benefits of voluntary work for mental well-being, including this one with 746 Swiss workers: Romualdo Ramos, Rebecca Brauchli, Georg Bauer, Theo Wehner, and Oliver Hammig, "Busy yet socially engaged: volunteering, work-life balance, and health in the working population," *Journal of Occupational and Environmental Medicine*, 2015, vol. 57, pp. 164–172.

"One study asked 105 German employees." (p. 117): Eva J. Mojza, Sabine Sonnentag, and Cladius Bornemann, "Volunteer work as a valuable leisure-time activity: A day-level study on volunteer work, non-work experiences, and well-being at work," *Journal of Occupational and Organizational Psychology*, 2011, vol. 84, pp. 123–52.

"This state of being completely absorbed in a task . . ." (p. 118): The Hungarian-American psychologist, Mihaly Csikszentmihalyi developed the concept of *flow* and conducted many studies to show that this psychological state is at the heart of thriving, productivity, and happiness. His seminal work is described in his book *Flow: The Psychology of Optimal Experience*, Harper and Row, 1990.

"This was shown in one study . . ." (p. 119): Kennon Sheldon, Robert Cummins, and Shanmukh Kamble, "Life balance and well-being: A novel conceptual and measurement approach," *Journal of Personality*, 2010, vol. 78, pp. 1093–133.

"In the table below there are 10 areas of life." (p. 120): This table is adapted from the Balance Assessment Sheet used in Study 1 from Kennon Sheldon, Robert Cummins, and Shanmukh Kamble, "Life balance and well-being: A novel conceptual and measurement approach." *Journal of Personality*, 2010, vol. 78, pp. 1093–133.

"In a survey conducted by the Mental Health Foundation in the UK in 2014 . . ." (p. 121): This report can be downloaded from the Mental Health Foundation website at https://www.mentalhealth.org.uk/a-to-z/w/work-life-balance.

"Instead, he advised the president to respond . . ." (p. 124): This incident and many others are discussed in a biography of Thompson written by his two daughters, Jenny Thompson and Sherry Thompson, *The Kremlinologist: Llewellyn E. Thompson. America's; Man in Cold War Moscow*, John Hopkins University Press, 2018.

"Too much empathy can lead to preferential treatment of others . . ." (p. 125): Adam Galinsky et al., "Why it pays to get inside the head of your opponent: The differential effects of perspective taking and empathy in negotiation," *Psychological Science*, 2008, vol. 19, pp. 378–84.

"To find out where you sit on the perspective-taking and empathy dials . . ." (p. 125): These questions have been adapted from the "Interpersonal

Reactivity Index": M. H. Davis, "A multidimensional approach to individual differences in empathy," *JSAS Catalog of Selected Documents in Psychology*, 1980, vol. 10, no. 85.

"It may surprise you to know that most of us are optimistic . . ." (p. 128): The science behind the "optimism bias" is explained in an engaging way by the University College London neuroscientist and psychologist Tali Sharot in her book *The Optimism Bias: Why We're Wired To Look On The Bright Side*, Robinson, 2012.

"Even during the coronavirus pandemic . . ." (p. 128): Laura Globig, Bastien Blain, and Tali Sharot, "When private optimism meets public despair: Dissociable effects on behavior and well-being." Paper under review but a pre-print can be accessed here: https://psyarxiv.com/gbdn8.

"Remember that optimism is not necessarily unrealistic . . ." (p. 129): I wrote about the importance of balancing optimism and pessimism for navigating the challenges of life in a previous book, *Rainy Brain, Sunny Brain: The New Science of Optimism and Pessimism*, William Heineman, 2012.

"Practice mindfulness" (p. 129): Lots of useful suggestions of how to practice mindfulness can be found in a book by Mark Williams and Danny Penman, *Mindfulness: A Practical Guide to Finding Peace in a Frantic World*, Piatkus Books, 2011.

"Studies have shown that generating aspects of a novel internally . . ." (p. 130): Raymond Mar et al., "Bookworms versus nerds: Exposure to fiction versus nonfiction, divergent associations with social ability, and the simulation of fictional social worlds," *Journal of Research in Personality*, 2006, vol. 40, pp. 694–712, and Matthijs Bal and Martijn Veltkamp, "How does fiction reading influence empathy: An experimental investigation on the role of emotional transportation," *PLOS One*, 2013, vol. 8, article e55341.

"It has been called the 'mind's flight simulator' . . ." (p. 130): Keith Oatley, "Fiction: Simulation of social worlds," *Trends in Cognitive Sciences*, 2016, vol. 20, pp. 618–28.

"Executive functions underpin our mental competence . . ." (p. 131): A good overview of executive functions and how they develop is available in a review by developmental psychologist Adele Diamond, "Executive functions," *Annual Review of Psychology*, 2013, vol. 64, pp. 135–68.

"In fact, studies with children have shown . . ." (p. 133): Terrie Moffitt et al., "A gradient of childhood self-control predicts health, wealth, and public safety," *Proceedings of the National Academy of Sciences,* 2011, vol. 108, pp. 2693–98.

"The results were encouraging." (p. 135): Our studies were conducted in collaboration with my colleague and friend Nazinin Derakshan, who is a cognitive psychologist based at Birkbeck College London and used what is called the "n-back task" to help high worriers gain more control over negative thinking. Our results are published in two scientific papers: Matthew Hotton, Nazanin Derskshan, and Elaine Fox, "A randomised controlled trial investigating the benefits of adaptive working memory training for working memory capacity and attentional control in high worriers," *Behavior Research and Therapy,* 2018, vol. 100, pp. 67–77, and Maud Grol et al., "The worrying mind in control: An investigation of adaptive working memory training and cognitive bias modification in worry-prone individuals," *Behavior Research and Therapy,* 2018, vol. 103, pp. 1–11.

"In one study, we selected volunteers who were binge eaters . . ." (p. 135): Danna Oomen, Maud Grol, Desiree Spronk, Charlotte Booth, and Elaine Fox, "Beating uncontrolled eating: Training inhibitory control to reduce food intake and food cue sensitivity," *Appetite,* 2018, vol. 131, pp. 73–83.

"One study in the Netherlands, for instance . . ." (p. 136): Artur Jaschke, Henkjan Honing, and Erik Scherder, "Longitudinal analysis of music education on executive functions in primary school children," *Frontiers in Neuroscience,* 2018, vol. 12, article 103.

Chapter 8: Know Thyself

"Our personality reflects our fundamental habits . . ." (p. 142): An excellent overview of the science of personality can be found in a book by psychologist and writer Christian Jarrett, *Be Who You Want: Unlocking the Science of Personality Change,* Robinson, 2021.

"This understanding has been called 'the psychology of the stranger' . . ." (p. 143): The American psychologist Dan McAdams coined the term "psychology of the stranger" to refer to the understanding of a person, or yourself, at the level of personality traits: Dan McAdams, "Personality, modernity, and the storied self: A contemporary framework for studying persons," *Psychological Inquiry,* 1996, vol. 7, pp. 295–321.

"Many people don't realize that the Myers-Briggs test was developed . . ."

(p. 144): The Myers-Briggs test was inspired by the ideas of Swiss psycho-analyst Carl Jung, who believed that each of us is born with four archetypes: the *Persona*, which is how we present ourselves to the world, the *Shadow*, our basic sexual desires and other life instincts, the *Anima* or *Animus*, which is the male and female version of our "true self," and the *Self*, which represents the entire conscious and unconscious aspects of a person. You can learn more about the life and ideas of Jung in a comprehensive biography by Deirdre Bair simply called *Jung: A Biography*, Little Brown, 2003.

"The test has many problems . . ." (p. 144): Merve Emre, who is a professor of English at the University of Oxford, has written a wonderful book about the strange history of the Myers-Briggs personality test: *What's Your Type: The Story of the Myers-Briggs, and How Personality Testing Took Over the World*, William Collins Publishers, 2018.

"And yet, people love taking this test . . ." (p. 144): In her book *What's Your Type*, Merve Emre describes the test as a "portal to an elaborate practice of talking and thinking about who you are," and we all love that.

"The more recent consensus from decades of scientific research . . ." (p. 144): Daniel Nettle, professor of anthropology at the University of Newcastle in the UK, has written a comprehensive and accessible overview of the Big Five dimensions of personality in his book *Personality: What Makes You the Way You Are*, Oxford University Press, 2009. Also see Christian Jarrett, *Be Who You Want: Unlocking the Science of Personality Change*, Robinson, 2021.

"To get an indication of how you fare . . ." (p. 145). This personality test was presented by Sam Gosling et al. (2003). A very brief measure of the Big Five Personality Domains. *Journal of Research in Personality*, 37, 504-528.

"*Conscientiousness* reflects your tendency to be diligent . . ." (p. 146): You can find out more about the science of conscientiousness in the bestselling book by psychologist Angela Duckworth, *Grit: The Power of Passion and Perseverance*, Vermilion, 2016.

"If you are introverted . . ." (p. 146): The impact for your life of where you sit on the introversion-extraversion spectrum is illustrated beautifully in the bestselling book by Susan Cain, *Quiet: The Power of Introverts in a World That Can't Stop Talking*, Viking, 2012.

"It's important to remember, however, that these traits can be modified . . ." (p. 148): To find lots of further evidence for this have a look at Christian Jarrett, *Be Who You Want: Unlocking the Science of Personality Change*, Robinson, 2021.

"Studies have confirmed . . ." (p. 148): Elizabeth Krumrei-Mancuso et al., "Links between intellectual humility and acquiring knowledge," *Journal of Positive Psychology*, 2020, vol. 15, pp. 155–70.

"For example, a 2018 survey found . . ." (p. 148): This survey was conducted by journalist Shane Snow and described in his engaging book *Dream Teams: Working Together Without Falling Apart*, Portfolio, 2018.

"Those who are willing to admit that they might be wrong are often happier . . ." (p. 149): You can find a good overview of the impact of humility on happiness and well-being from several different authors in a book edited by American psychologist Jennifer Cole Wright, *Humility*, Oxford University Press, 2019.

"Given the relative consistency of this habit of mind . . ." (p. 149): Michael Ashton et al., "A six-factor structure of personality-descriptive adjectives: solutions from psycholexical studies in seven languages," *Journal of Personality and Social Psychology*, 2004, vol. 86, pp. 356–66. An accessible overview of the Honesty-Humility dimension of personality, and its importance, is provided by Kibeom Lee and Michael Ashton in their book *The H Factor of Personality*, Wilfrid Laurier University Press, 2012.

"The well-known social psychologist John Bargh . . ." (p. 149): John Bargh, Mark Chen, and Lara Burrows, "Automaticity of social behavior: Direct effects of trait construct and stereotype activation on action," *Journal of Personality and Social Psychology*, 1996, vol. 71, pp. 230–44. John Bargh worked on many social priming studies over the years, providing fascinating material for bestselling books by, for example, Malcolm Gladwell, *Blink: The Power of Thinking Without Thinking*, Penguin, 2006, and Daniel Kahneman, *Thinking Fast and Slow*, Penguin, 2012. Bargh described this work in his own book *Before You Know It: The Unconscious Reasons We Do What We Do*, Windmill Books, 2017.

"Fast-forward to 2012 and a group of Brussels-based psychologists . . ." (p. 150): Stephane Doyen, Olivier Klein, Cora-Lise Pichon, and Axel Cleermans, "Behavioral priming: It's all in the mind, but whose mind?" *PLOS One*, January 18, 2021, article 0029081.

"Bargh was furious . . ." (p. 150): The science journalist Ed Yong wrote about John Bargh's extraordinary reaction to failures to replicate his work: "A failed replication draws a scathing personal attack from a psychology professor," *National Geographic*, March 2012.

"For example, in a fascinating line of research . . ." (p. 150): Leor Zmigrod et al., "The psychological roots of intellectual humility: The role of intelligence and cognitive flexibility," *Personality and Individual Differences*, 2019, vol. 141, pp. 200–8.

"To give you an idea of how you rate in terms of intellectual humility . . ." (p. 151): This scale was developed by psychologists Tenelle Porter and Karina Schumann, "Intellectual Humility and openness to the opposing view," *Self and Identity*, 2018, vol. 17, pp. 139–62.

"Our intellectual humility can be increased by nurturing a growth mindset . . ." (p. 152): This was reported in study 4 in Porter and Schumann, "Intellectual Humility and openness to the opposing view," *Self and Identity*, 2018, vol. 17, pp. 139–62. The remarkable impact that "fixed" and "growth" mindsets can have on a person's life is described beautifully by psychologist Carol Dweck in her book *Mindset: The New Psychology of Success*, Ballantine Books, 2007.

"Understanding these subtle signals from your body . . ." (p. 154): Interoception—the ability to describe our physical feelings and sensations—is now seen as critical to self-awareness in psychology. Excellent overviews can be found in a book by A. D. (Bud) Craig, *How Do You Feel? An Interoceptive Moment with Your Neurobiological Self*, Princeton University Press, 2014, and in a book by Guy Claxton called *Intelligence in the Flesh: Why Your Mind Needs Your Body Much More Than It Thinks*, Yale University Press, 2015.

"At the dawn of scientific psychology in the USA in 1884 . . ." (p. 154): A description of James's theory can be found in an article he wrote in 1884, "*What Is an Emotion?*" It is reprinted in the book *Heart of William James*, edited by Robert Richardson, Harvard University Press, 2012.

"One technique is called the 'heartbeat detection task' . . ." (p. 156): While this task is still widely used and can help us to tune in to our internal signals, it does have problems: Georgia Zamariola et al., "Interoceptive accuracy scores from the heartbeat counting task are problematic: Evidence from simple bivariate correlations," *Biological Psychology*, 2018, vol. 137, pp. 12–17.

"In general, however, our own insight into our internal sensations . . ." (p. 156): The work of Sarah Garfinkel and Hugo Critchley at the University of Sussex has shown that people's perception of how good they are at reading their internal signals is not actually very accurate. A good overview of this work can be found in a *Wired* article written by Joao Medeiros, "Listening to your heart might be the key to conquering anxiety," October 20, 2020, https://www.wired.co.uk/article/sarah-garfinkel-interoception.

"While self-report measures are not ideal either . . ." (p. 156): These questions are adapted from various questionnaires that are widely used in research. These include the *Body Perception Questionnaire* developed by Stephen Porges: https://www.stephenporges.com/body-scales, and the *Multi-dimensional Assessment of Awareness Scale*, Wolf Mehlings and colleagues, "The multidimensional assessment of interoceptive awareness," *PLOS One*, 2012, vol. 7, article e48230.

"And this setup seems to be important in helping us to differentiate . . ." (p. 157): A compelling case for the role of interoception in forming self-awareness is made by Antonio Damasio in *Self Comes to Mind: Constructing the Conscious Brain*, Vintage, 2012.

"We know this because . . ." (p. 157): Mariana Babo-Rebelo, Craig Richter, and Catherine Tallon-Baudry, "Neural responses to heartbeats in the default network encode the self in spontaneous thoughts," *Journal of Neuroscience*, 2016, vol. 36, pp. 7829–40.

"Studies like this tell us that the mind is best understood . . ." (p. 157): Babo-Rebelo, Richter, and Tallon-Baudry, see above. Interesting overviews are also available in Damiano Azzalini, Ignacio Rebollo, and Catherine Tallon-Baudry, "Visceral signals shape brain dynamics and cognition," *Trends in Cognitive Sciences*, 2019, vol. 23, pp. 488–509, and A. D. (Bud) Craig, "How do you feel—now?" *Nature Reviews Neuroscience*, 2009, vol. 10, pp. 59–70.

"This point takes on a particular significance . . ." (p. 158): This work is beautifully described in a book by Boston-based psychologist and neuroscientist, Lisa Feldman Barrett, *Seven and a Half Lessons About the Brain*, Picador, 2020.

"More importantly, these signals have even been shown . . ." (p. 158): Ruben Azevedo, Sarah Garfinkel, Hugo Critchley, and Manos Tsakiris, "Cardiac afferent activity modulates the expression of racial stereotypes," *Nature Communications*, 2017, article 13854.

"In the USA, Black people are more than twice as likely to be unarmed when killed . . ." (p. 158): This is a shocking statistic that has been found in a detailed analysis of data from the USA: Cody Ross, "A multi-level Bayesian analysis of racial bias in police shootings at the county-level in the United States," *PLOS One*, vol. 10, e0141854.

"Potential reasons for this depressing statistic . . ." (p. 158): Joshua Correll, Bernadette Park, and Charles Judd, "The police officer's dilemma: Using ethnicity to disambiguate potentially threatening individuals," *Journal of Personality and Social Psychology*, 2002, vol. 83, pp. 1314–29.

"Further research has shown us that most of the misidentifications. . . ." (p. 159): Ruben Azevedo, Sarah Garfinkel, Hugo Critchley, and Manos Tsakiris, "Cardiac afferent activity modulates the expression of racial stereotypes," *Nature Communications*, 2017, article 13854.

"Due to unconscious bias, Black men ..." (p. 159): John Paul Wilson, Kurt Hugenberg, and Nicholas O. Rule, "Racial bias in judgments of physical size and formidability: From size to threat," *Journal of Personality and Social Psychology*, 2017, vol. 113, pp. 59–80.

"There are several studies that show ..." (p. 160): Robert Schwitzgebel, "The performance of Dutch and Zulu adults on selected perceptual tasks," *The Journal of Social Psychology*, 1962, vol. 57, pp. 73–77, and Darhl Pedersen and John Wheeler, "The Müller-Lyer illusion among Navajos," *The Journal of Social Psychology*, 1982, vol. 121, pp. 3–6.

"An intriguing study with financial traders . . ." (p. 161): Narayanan Kandasamy et al., "Interoceptive ability predicts survival on a London trading floor," *Scientific Reports*, 2016, vol. 6, article 32986.

"One team of researchers set out to look at the impact of meditation . . ." (p. 162): Laura Mirams, Ellen Poliakoff, Richard Brown, and Donna Lloyd, "Brief body-scan meditation practice improves somatosensory perceptual decision making," *Consciousness and Cognition*, 2013, vol. 22, pp. 348–59.

Chapter 9: Beliefs and Values

"Many of the spectators thought it was a joke . . ." (p. 165): There is a good overview of Cliff Young's story in a blog by Darko Kankaras published in "Monitor the Beat": https://monitorthebeat.com/blogs/news/cliff -young-the-legend-of-ultramarathon.

"To uncover a core belief yourself . . ." (p. 168): There is a lot of very helpful advice on the web platform PositivePsychology.com, which provides evidence-based tips from practitioners. You can find several helpful work-sheets to help you identify and challenge your core beliefs here: https:// positivepsychology.com/core-beliefs-worksheets.

"In fact, it's only when you identify and understand your core values . . ." (p. 170): A highly accessible overview of how to identify your deepest values can be found in a book by Australian therapist and life coach Russ Harris, *The Happiness Trap: Stop Struggling, Start Living*, Robinson Books, 2008.

"Our political beliefs are among the most difficult to change . . ." (p. 172): Studies using brain imaging have found that when one of our strongly held beliefs is challenged, there is more activation in the parts of the brain that are associated with self-identity and strong emotions: Jonas Kaplan, Sarah Gimbel, and Sam Harris, "Neural correlates of maintaining one's political beliefs in the face of counterevidence," *Scientific Reports*, 2016, vol. 6, article 39589.

"This is 'confirmation bias' . . ." (p. 173): Raymond Nickerson, "Confirmation bias: A ubiquitous phenomenon in many guises," *Review of General Psychology*, 1998, vol. 2, pp. 175–220.

"In a quirk of mind called 'cognitive dissonance' . . ." (p. 173): The classic example of how we often lean into a cherished belief rather than change it when it is challenged occurred back in 1954 in Illinois when a group of people who believed they were going to be rescued from a flood by aliens called the "Guardians" waited all night. When no flood occurred and no alien ship came to rescue them, rather than shifting their belief they concluded that they must have got the date wrong and doubled down on their belief in the alien rescue. Three of the researchers who investigated this dramatic example of cognitive dissonance wrote about it: Leon Festinger, Henry Riecken, and Stanley Schachter, *When Prophecy Fails*, Pinter & Martin, reprint 1958.

"Take the case of PJ Howard . . ." (p. 174): The story of PJ Howard and his partner, Sharon Collins, is told in several Irish newspapers from the time including an article by Emer Connolly, "He was still in love with her, and desperately wanted to believe her," *Irish Independent*, July 13, 2008.

"We are essentially 'cognitive misers' . . ." (p. 175): This term was coined by the Scottish psychologist Colin Macrae, who has conducted several studies showing that stereotypes and other beliefs can free up mental resources: Colin Macrae, Alan Milne, and Galen Bodenhausen, "Stereotypes as energy-saving devices: A peek inside the cognitive toolbox," *Journal of Personality and Social Psychology*, 1994, vol. 66, pp. 37–47.

"We often become reliant on our more superficial selves . . ." (p. 177): The French philosopher Jean-Paul Sartre spoke about this in terms of what he called "bad faith." Jean-Paul Sartre, *Essays in Existentialism*, Citadel Press, 1993, pp. 167–69.

"Our personal stories create meaning for us . . ." (p. 178): Dan McAdams, a psychologist at Northwestern University in the US, has spent most of his professional life investigating how we make sense of our true self by means of our life stories. You can find a good overview of his and others research in his book *The Art and Science of Personality Development*, Guilford Press, 2015. Journalist Julie Beck has also written a highly accessible overview of this work in *The Atlantic* magazine in a 2015 article, "Story of My Life: How Narrative Creates Personality."

"According to research by developmental psychologists . . ." (p. 180): Kate McLean, Monisha Pasupathi, and Jennifer Pals, "Selves creating stories creating selves: A process model of self-development," *Personality and Social Psychology Review*, 2007, vol. 11, pp. 262–78.

"The stories we tell ourselves matter hugely." (p. 180): A team of psychologists conducted a mega-analysis to find the typical structures of the stories we tell about ourselves: Kate McLean et al., "The empirical structure of narrative identity: The initial Big Three," *Journal of Personality and Social Psychology*, 2020, vol. 119, pp. 920–44.

"The science tells us that stories of redemption . . ." (p. 180): A good overview of this line of work can be found in Dan McAdams, *The Redemptive Self: Stories Americans Live By*, Oxford University Press, 2006.

Chapter 10: Understanding Your Emotions

"Two components of successful influence stood out." (p. 189): Kevin Dutton, *Flipnosis: The Art of Split-Second Behavior*, William Heinemann, 2010. For more on the evolutionary principles of persuasion, see Robert B. Cialdini, *Influence: The Psychology of Persuasion*, Harper Business, revised edition, 2006.

"For instance, we now know that anger can be a very effective negotiating tool . . ." (p. 190): A series of fascinating studies from the University of Amsterdam psychologist Gerben Van Kleef show us how powerful anger can be in negotiations: Gerben Van Kleef and Stephane Cote, "Expressing anger in conflict: When it helps and when it hurts," *Journal of Applied Psychology*, 2007, vol. 92, pp. 1557–69.

"Angry buyers are more likely to get a better deal . . ." (p. 190): Gerben Van Kleef, Carsten De Dreu, and Antony Manstead, "The interpersonal effects of anger and happiness in negotiations," *Journal of Personality and Social Psychology*, 2004, vol. 86, pp. 57–76.

"Expressing anger toward those with much more power . . ." (p. 190): Gerben Van Kleef, Carsten De Dreu, Davide Pietroni, and Antony Manstead,

"Power and emotion in negotiation: Power moderates the interpersonal effects of anger and happiness on concession making," *European Journal of Social Psychology*, 2006, vol. 36, pp. 557–81.

"Emotions play a crucial role in helping us adapt to change . . ." (p. 191): Psychologists Keith Oatley and Philip Johnson-Laird developed a cognitive theory of emotions that introduced the idea of emotions as levers, which helps us to move between goals, back in 1987: "Toward a cognitive theory of emotions," *Cognition and Emotion*, 1987, vol. 1, pp. 29–50.

"There are two broad schools of thought in affective science . . ." (p. 193): I wrote an overview, and possible compromise, between the two schools of thought in affective science about how emotions are made—the classical and the constructed emotion perspectives—in the following scientific paper: Elaine Fox, "Perspectives from affective science on understanding the nature of emotion," *Brain and Neuroscience Advances*, 2018, vol. 2, pp. 1–8.

"What has been called the 'classical view' . . ." (p. 193): There are many examples of this perspective. A comprehensive overview can be found in an academic book by neuroscientist Jaak Panksepp, which has a heavy emphasis on animal research: *Affective Neuroscience*, Oxford University Press, 1998. Overviews of work on human participants can be found in articles by Paul Ekman and Daniel Cordaro, "What is Meant by Calling Emotions Basic?" *Emotion Review*, 2011, vol. 3, pp. 364–70, and Caroll Izard, "Basic emotions, natural kinds, emotion schemas, and a new paradigm," *Perspectives on Psychological Science*, 2007, vol. 2, pp. 270–80. A comprehensive overview of this view of basic emotions being examples of what philosophers call "natural kinds" can be found in Lisa Feldman Barrett, "Are emotions natural kinds?" *Perspectives on Psychological Science*, 2006, vol. 1, pp. 28–58.

"In the 1960s, a popular idea in psychology . . ." (p. 193): The idea of a triune brain, which evolved layer by layer, was first developed by the pioneering American neuroscientist Paul MacLean. His theory is well explained in his 1990 book *The Triune Brain in Evolution: Role in Paleocerebral Functions*, Plenum Press.

"While there is some degree of structural truth . . ." (p. 193): While the triune brain idea is no longer taken seriously in neuroscience, largely because we now know that brain systems were not "added on" over the course of evolution as assumed by this theory, we should not forget that the work of Paul MacLean was an important forerunner to an evolutionary view of the brain, which is still very much in vogue.

"Instead, just like an organization . . ." (p. 195): This metaphor was used by psychologist and neuroscientist Lisa Feldman Barrett in her highly readable book *How Emotions are Made*, Macmillan, 2017. This book sets out a very different perspective to the classical view of emotions and argues that, rather than being born with prewired emotion circuits built into our brain, our emotional life is instead largely *constructed* throughout our lives. This perspective draws on a much older line of work that shows that how we interpret physiological arousal *and* our social surroundings ultimately determines how we feel. These seminal studies were conducted by psychologists Stanley Schachter and Jerome Singer, "Cognitive, social, and physiological determinants of emotional state," *Psychological Review*, 1962, vol. 69, pp. 379–99.

"Instead, what people do describe are much broader *dimensions* . . ." (p. 196): A dimensional model of emotion was first proposed by the founder of scientific psychology Wilhelm Wundt in his *Outlines of Psychology*, originally published in 1897. An influential version of this perspective is called the "circumplex model" and was developed by the American psychologist James Russell. Good academic overviews are available in: James Russell, "A circumplex model of affect," *Journal of Personality and Social Psychology*, 1980, vol. 39, pp. 1161–78, and Jonathan Posner, James Russell, and Bradley Peterson, "The circumplex model of affect: An integrative approach to affective neuroscience, cognitive development, and psychopathology," *Developmental Psychopathology*, 2005, vol. 17, pp. 715–34.

"The results confirm that the brain operates as a highly fluid connected system . . ." (p. 197): There is a highly readable overview of how the brain works in a book by the Boston-based psychologist and neuroscientist Lisa Feldman Barrett: *Seven and a Half Lessons About the Brain*, Picador, 2020.

"What this tells us is that changes in the body are transformed into an emotion . . ." (p. 198): This theoretical perspective has been put forward in a compelling book by Lisa Feldman Barrett, *How Emotions Are Made: The Secret Life of the Brain*, Macmillan, 2017.

"These observations force us to reconsider . . ." (p. 199): See Lisa Feldman Barrett's book *How Emotions are Made* for much more detail on this constructed emotion perspective.

"We know that people are typically able to remember around seven items . . ." (p. 200): George Miller, "The magical number seven, plus or minus two: Some limits on our capacity for processing information," *Psychological Review*, 1956, vol. 63, pp. 81–97.

"Bodily feelings conspire with powerful thoughts to elicit action . . ." (p. 201):
This perspective is described beautifully by psychologist and neuroscientist Lisa Feldman Barrett in her compelling book *How Emotions are Made*, Macmillan, 2017.

"Positive emotions tend to widen our attention . . ." (p. 203): Barbara Fredrickson and Christine Branigan, "Positive emotions broaden the scope of attention and thought-action repertoires," *Cognition and Emotion*, 2005, vol. 19, pp. 313–32. The work of Barbara Fredrickson and her team has provided us with a blueprint for understanding how positive emotions play a role in health, well-being, and the building of resilience with her "broaden and build" theory. A great overview of her work can be found in two of her inspirational books: *Positivity*, Crown Archetype Publishers, 2009, and *Love 2.0*, Plume, 2013.

"This reward center can be divided into two parts . . ." (p. 204): Kent Berridge and Terry Robinon, "Parsing reward," *Trends in Neuroscience*, 2003, vol. 26, pp. 507–13.

"whereas the 'wanting' parts release the chemical dopamine . . ." (p. 204): We know about the inner workings of the reward system to a large extent because of the detailed research of the University of Michigan neuroscientist and psychologist Kent Berridge. A good overview of his discoveries about dopamine can be found in Kent Berridge, "Affective valence in the brain: Modules or modes?" *Nature Reviews Neuroscience*, 2019, vol. 20, pp. 225–34.

"When you are in a good mood . . ." (p. 204): The work of American psychologist Alice Isen has been foundational in showing us how powerful simple positive events can be in terms of boosting our creativity, enhancing our resilience, and improving our decision-making. Some key papers are: Alice Isen et al., "The influence of positive affect on clinical problem solving," *Medical Decision Making*, 1991, vol. 11, pp. 221–27; Alice Isen and colleagues, "Positive affect facilitates creative problem solving," *Journal of Personality and Social Psychology*, 1987, vol. 51, pp. 1122–31; and Gregory Ashby, Alice Isen, and And Turken, "A neuropsychological theory of positive affect and its influence on cognition," *Psychological Review*, 1999, vol. 106, pp. 529–50.

"And even experienced doctors . . ." (p. 204): Carlos Estrada et al., "Positive affect improves creative problem solving and influences reported source of practice satisfaction in physicians," *Motivation and Emotion*, 1994, vol. 18, pp. 285–99.

"Positive emotional experiences can guide all of us . . ." (p. 205): A comprehensive overview of the science behind emotions and decision-making can be found in the following academic review: Jennifer Lerner et al., "Emotion and decision making," *Annual Review of Psychology*, 2015, vol. 66, pp. 799–823.

"After the terrorist attacks of 9/11 . . ." (p. 205): Barbara Fredrickson et al., "What good are positive emotions in crisis: A prospective study of resilience and emotions following the terrorist attacks on the United States on September 11th, 2001," *Journal of Personality and Social Psychology*, 2003, vol. 84, pp. 365–76.

"Positive experiences and emotions can also be 'banked' . . ." (p. 205): Barbara Fredrickson and Robert Levenson, "Positive emotions speed recovery from the cardiovascular sequelae of negative emotions," *Cognition and Emotion*, 1998, vol. 12, pp. 191–220.

"Years of painstaking research have shown . . ." (p. 205): The University of North Carolina psychologist Barbara Fredrickson discovered that in order to thrive you need to experience at least three positive events for every one negative event. You can find out more about this research and how to boost your positivity ratio in her book *Positivity*, and you can also check your own positivity ratio on her website: http://www.positivityratio.com /index.php.

Chapter 11: Learning to Regulate Your Emotions

"We can learn a lot about how to regulate strong emotions . . ." (p. 210): A brief overview of dialectical behavioral therapy (DBT) and its origins can be found in a paper by Linda Dimeff and Marsha Linehan, "Dialectical Behavior Therapy in a Nutshell," *The California Psychologist*, 2001, vol. 34, pp. 10–13.

"Dialectical behavioral therapy (DBT) suggests . . ." (p. 211): The following technical manual provides numerous exercises to help you regulate your emotions. Matthew McKay and Jeffrey C. Wood, *The Dialectical Behavior Skills Workbook: Practical DBT exercises for learning mindfulness, interpersonal effectiveness, emotion regulation and distress tolerance*, New Harninger, 2nd edition, 2019.

"Thankfully, there are many actions you can take to regulate specific emotions." (p. 212): Stanford psychologist James Gross has led the way in studying emotion regulation and has developed what he calls the "process" theory of emotion regulation. This framework is very influential and has inspired a mountain of research on emotion regulation. The idea that emotions can be regulated at different time points was investigated in James Gross, "Antecedent- and response-focused emotion regulation: Divergent consequences for experience, expression, and physiology," *Journal of Personality and Social Psychology*, 1998, vol. 74, pp. 224–37. An excellent overview of research and theory in emotion regulation with chapters from various experts can be found in *The Handbook of Emotion Regulation*, edited by James Gross, Guilford Press, 2nd edition, 2015.

"We still know surprisingly little . . ." (p. 213): Meghann Matthews, Thomas L. Webb, Roni Shafir, Miranda Snow, and Gal Sheppes, "Identifying the determinants of emotion regulation: a systematic review with meta-analysis," *Cognition and Emotion*, published online June 24, 2021.

"The diagram below illustrates the four general types of strategies . . ." (p. 214): This diagram is an adaptation of the process model of emotion regulation as outlined by James Gross and colleagues. Original versions can be seen in Fig. 1 of Kateri McRae and James Gross, "Introduction to a special issue on Fundamental Questions in Emotion Regulation,'" *Emotion*, 2020, vol. 20, pp. 1–9.

"Our brain will always magnify potential danger more than potential reward . . ." (p. 217): I wrote extensively about this research and its implications in a previous book: Elaine Fox, *Rainy Brain, Sunny Brain: The New Science of Optimism and Pessimism*, William Heineman, 2012.

"The problem arises when these thoughts become a habitual response . . ." (p. 218): In psychology, "negative automatic thoughts" were identified by Aaron Beck, who is widely considered the "father of cognitive behavioral therapy," better known as talking therapy. You can access many useful resources on the Beck Institute's website: https://beckinstitute.org/resources-for-professionals/multimedia-resources. The handy acronym ANT (automatic negative thoughts) is frequently used.

"So, an important aspect of emotion regulation . . ." (p. 218): The University of Michigan psychologist and affective scientist Ethan Kross has written a wonderful book about the nature of "chatter" and how you can refocus and reframe the negative chatter in your head and have a happier life: *Chatter: The Voice in Our Head and How to Harness It*, Vermilion Press, 2021.

"Clinical psychologists find that this simple technique works wonders." (p. 220): My Oxford colleague, clinical psychologist Jennifer Wild, uses these techniques extensively to help people rebuild their lives after major trauma. She has written an engaging and informative book about how to use these techniques in your own life: *Be Extraordinary: Seven Key Skills to Transform Your Life from Ordinary to Extraordinary*, Robinson, 2020.

"Indeed, some studies have found that it is the flexibility . . ." (p. 222): The Columbia University psychologist George Bonanno has conducted seminal work on emotion regulation flexibility, summarized in George Bonanno and Charles Burton, "Regulatory flexibility: An individual differences perspective on coping and emotion regulation," *Perspectives on Psychological Science*, 2013, vol. 8, pp. 591–612. Other interesting overviews and thoughts on when to choose different regulatory strategies can be found in James Gross, "Emotion regulation: Current status and future prospects," *Psychological Inquiry*, 2015, vol. 26, pp. 1–26, and Ethan Kross, "Emotion regulation growth points: Three more to consider," *Psychological Inquiry*, 2015, vol. 26, pp. 69–71.

"For instance, one study was conducted . . ." (p. 222): George Bonanno et al., "The importance of being flexible: The ability to both enhance and suppress emotional expression predicts long-term adjustment," *Psychological Science*, 2004, vol. 15, pp. 482–87.

"One powerful therapeutic approach . . ." (p. 223): There are several books that provide a great overview of ACT. Two I have found especially helpful: one written by the founder of ACT Steven Hayes, *A Liberated Mind: The Essential Guide to ACT*, Vermillion Press, 2019, and a highly accessible self-help book by Russ Harris, *The Happiness Trap*, Robinson Publishing, 2008.

"One study involved a group of computer engineers . . ." (p. 225): Stephanie Spera, Eric Buhrfeind, and James Pennebaker, "Expressive writing and coping with job loss," *Academy of Management Journal*, vol. 37, pp. 722–33.

"So stick with it, because the benefits of expressive writing . . ." (p. 226): You can read an overview of this work by the leading psychologist in the field, James Pennebaker, in his entertaining book *The Secret Life of Pronouns*, Bloomsbury Publishing, 2013.

"One study has shown that the key to psychological health . . ." (p. 226): In a fascinating series of studies, a team of psychologists from the Universities of Toronto and Berkeley, California, found that accepting negative emotions has a range of benefits for our psychological health: Brett Ford, Phoebe Lam, Oliver John, and Iris Mauss, "The psychological health benefits of accepting negative emotions and thoughts: Laboratory, diary, and longitudinal evidence," *Journal of Personality and Social Psychology*, 2018, vol. 115, pp. 1075–92.

"There are plenty of techniques to choose from." (p. 228): Many techniques that help to quell your emotions come from mindfulness practice. Much has been written about mindfulness and many apps and books are available. A book by my Oxford colleague, psychologist Mark Williams, that he wrote with journalist Danny Penman is particularly useful: *Mindfulness: A Practical Guide to Finding Peace in a Frantic World*, Little Brown Books, 2011. Another great book is by comedian Ruby Wax, *A Mindfulness Guide for the Frazzled*, Penguin Life, 2016.

"Most hostage negotiators . . ." (p. 229): This is called the "Behavior Change" Stairway Model and was developed by Gary Noesner, the former head of the FBI's hostage negotiation unit. You can read more about his life and work in his book *Stalling for Time: My Life as an FBI Hostage Negotiator*, Fodor Travel Publications, 2010.

"When asked to list the top attributes of a good negotiator . . ." (p. 231): Kirsten Johnson, Jeff Thomson, Judith Hall, and Cord Meyer, "Crisis (hostage) negotiators weigh in: The skills, behaviors, and qualities that characterize an expert crisis negotiator," *Police Practice and Research*, 2018, vol. 19, pp. 472–89.

"The capacity to describe emotional feelings in fine-grained detail . . ." (p. 232): A good overview can be found in Todd Kashdan, Lisa Feldman Barrett, and Patrick McKnight, "Unpacking emotion differentiation: Transforming unpleasant experience by perceiving distinctions in negativity," *Current Directions in Psychological Science*, 2015, vol. 24, pp. 10–16, as well as in *How Emotions are Made* by Lisa Feldman Barrett.

"The power of emotional granularity . . ." (p. 232): Lisa Feldman Barrett et al., "Knowing what you're feeling and knowing what to do about it: Mapping the relation between emotion differentiation and emotion regulation," *Cognition and Emotion*, 2001, vol. 15, pp. 713–24.

"Labeling our positive feelings . . ." (p. 233): Michelle Tugade, Barbara Fredrickson, and Lisa Feldman Barrett, "Psychological resilience and positive emotional granularity: Examining the benefits of positive emotions on coping and health," *Journal of Personality*, 2004, vol. 72, pp. 1161–90.

Chapter 12: The Nature of Intuition

"Many studies in psychology tell us that intuition . . ." (p. 239): Several highly accessible books on the power of intuition and our unconscious mind in our everyday lives are available. A provocative and engaging example is *Blink: The Power of Thinking Without Thinking*, Penguin Books, 2006, by science writer Malcolm Gladwell. Another good read is *The Hidden Brain*, Penguin Books, 2010, by journalist Shankar Vedantam.

"This was demonstrated in a now classic study . . ." (p. 239): Antoine Bechara, Hanna Damasio, Daniel Tranel, and Antonio Damasio, "Deciding advantageously before knowing the advantageous strategy," *Science*, 1997, vol. 275 (5304), pp. 1293–95.

"Intuition is that part of our mind that presents us . . ." (p. 239): The foundational work on how we develop intuitive knowledge about complex environments was conducted by the City University of New York psychologist Arthur Reber and called "implicit learning": Arthur Reber, "Implicit learning and tacit knowledge," *Journal of Experimental Psychology: General*, 1989, vol. 118, pp. 219–35. An academic overview of work in this area can be found in a book edited by Alex Cleermans, Viktor Allakhverdov, and Maria Kuvaldina, *Implicit Learning: 50 Years On*, Routledge, 2019.

"It's the 'gut feeling' that's easily missed." (p. 239): A great overview of intuition can be found in a book by the German psychologist Gerd Gigerenzer, *Gut Feelings: Short Cuts to Better Decision Making*, Penguin, 2008.

"This ability to deduce vital information . . ." (p. 241): The term "thin slicing" was coined by Stanford University social psychologist Nalini Ambady and was popularized by the science writer Malcolm Gladwell in his bestselling book *Blink: The Power of Thinking Without Thinking*, Penguin, 2006.

"In one well-known study, students were asked to evaluate their professors . . ." (p. 241): Nalini Ambady and Robert Rosenthal, "Half a minute: Predicting teacher evaluations from thin slices of nonverbal behavior and physical attractiveness," *Journal of Personality and Social Psychology*, 1993, vol. 64, pp. 431–41.

"It is based on what is called 'tacit knowledge' . . ." (p. 242): The concept of tacit knowledge was introduced by chemist turned philosopher of science Michael Polanyi in his book *Personal Knowledge*, Routledge, 1998 (first published 1958). He argued that this type of knowledge that comes through traditions, implied values, and inherited practices is often under-rated and is actually a crucial part of scientific practice.

"Donald Rumsfeld, the former US Secretary of Defense . . ." (p. 242): Donald Rumsfeld explained the limitations of intelligence reports in a White House briefing on February 12, 2002, by saying that "There are known knowns. There are things we know we know. We also know there are known unknowns. That is to say, we know there are some things we do not know. But there are also unknown unknowns, the ones we don't know we don't know."

"My own work on the profound impact of danger signals . . ." (p. 242): Elaine Fox, "Processing emotional facial expressions: The role of anxiety and awareness," *Cognitive, Affective & Behavioral Neuroscience*, 2002, vol. 2, pp. 52–63.

"What was surprising was that when I prevented . . ." (p. 243): Blocking the perception of the faces was achieved by a technique called "backward masking," which I did by superimposing an image of a jumbled-up neutral face on top of the original image almost immediately—after just seventeen milliseconds, so that the original image could no longer be seen.

"Albert Einstein has been widely quoted . . ." (p. 243): It turns out that Einstein almost certainly did not say this quote, but did believe strongly in the power of the intuitive mind. You can read about the history of this quote here: https://quoteinvestigator.com/2013/09/18/intuitive-mind.

"Called the 'enteric nervous system' . . ." (p. 244): You can learn everything you ever wanted to know about your guts in an entertaining read by the German science writer Guilia Enders in her book *Gut: The Inside Story of our Body's Most Under-rated Organ*, Scribe, 2015. If you want a more academic overview, try Meenakshi Rao and Michael Gershon, "The bowel and beyond: The enteric nervous system in neurological disorders," *Nature Reviews Gastroenterology and Hepatology*, 2016, vol. 13, pp. 517–28.

"Of course, in a wider sense context becomes culture . . ." (p. 245): A compre-hensive overview of the impact of culture and context on practical decision-making and problem-solving can be found in a book edited by psychologist Robert Sternberg et al., *Practical Intelligence in Everyday Life*, Cambridge University Press, 2000.

"In a series of studies conducted in rural Kenya . . ." (p. 245): Robert Sternberg et al., "The relationship between academic and practical intelligence: A case study in Kenya," *Intelligence*, 2001, vol. 29, pp. 401–18.

"We can also see this demonstrated . . ." (p. 247): Stephen Ceci and Jeffrey

Liker, "A day at the races: A study of IQ, expertise, and cognitive complexity," *Journal of Experimental Psychology: General*, 1986, vol. 115, pp. 255–66.

"You might think that a person's ability . . ." (p. 248): Stephen Ceci and Ana Ruiz, "The role of general ability in cognitive complexity: A case study of expertise," in Robert Hoffman (ed.) *The Psychology of Expertise*, Springer, 1992.

"The value of intuition is often downplayed . . ." (p. 248): In their book *The ESP Executive*, Prentice Hall, 1974, business analysts Douglas Dean and John Mihalasky described several studies where thousands of executives admitted to regularly using their intuitions, which the authors referred to as extra-sensory perception (ESP), when making business decisions. A good overview of the importance of using intuition alongside more traditional rational analysis in business is provided by Eugene Sadler-Smith and Erella Shefy, "The intuitive executive: Understanding and applying 'gut feel' in decision-making," *Academy of Management Executive*, 2004, vol. 18, pp. 76–91.

"The cognitive scientist Herbert Simon . . ." (p. 249): Herbert Simon, "What is an explanation of behavior?" *Psychological Science*, 1992, vol. 3, pp. 150–61.

Chapter 13: Looking Outside: How Context Fuels Intuition

"In the 1950 Monaco Grand Prix . . ." (p. 251): I came across the story of Juan Manuel Fangio in a blog on ClientWise.com by Chris Holman: "Trusting Your Gut," April 22, 2010.

"Called 'context sensitivity' in psychology . . ." (p. 252): The Columbia University psychologist George Bonanno and his team have developed a scale to measure individual differences in context sensitivity: George Bonanno, Fiona Maccallum, Matteo Malgaroli, and Wai Kai Hou, "The Context Sensitivity Index (CSI): Measuring the ability to identify the presence and absence of stressor context cues," *Assessment*, 2020, vol. 27, pp. 261–73.

"I recently got a personal taste of context sensitivity . . ." (p. 252): Interest in psychics is at an all-time high; according to some YouGov reports, one-third of the UK population say they have sought help from a psychic, especially during periods of personal or political uncertainty. These visits tend to increase dramatically after major events such as the Boston Marathon bombing and the coronavirus pandemic.

"Doing rather than thinking . . ." (p. 254): We accumulate a vast data bank of knowledge across a lifetime that helps us to pick up on subtle cues and patterns in our environment: Peter Frensch and Dennis Runger, "Implicit Learning," *Current Directions in Psychological Science*, 2003, vol. 12, pp. 13–18.

"As C. S. Lewis himself reminds us . . ." (p. 255): Clive Staples Lewis, *Letters to Malcolm*, HarperOne, 2017 (originally published 1964).

"In 2004 Lieutenant Donovan Campbell . . ." (p. 256): Donovan Campbell, *Joker One: A Marine Platoon's Story of Courage, Leadership, and Brotherhood*, Presidio Press, 2010.

"My research has shown, for instance . . ." (p. 257): This was work conducted in collaboration with fellow cognitive psychologist, UCL-based Nilli Lavie. We ran several experiments showing that if you have to keep four or more things in mind, outside distractions have little effect: Nilli Lavie and Elaine Fox, "The role of perceptual load in negative priming," *Journal of Experimental Psychology: Human Perception and Performance*, 2000, vol. 26, pp. 1038–52.

"In one study, some volunteers were asked to steal $20 . . ." (p. 258): Andrea Webb et al., "Effectiveness of pupil diameter in a probable-lie comparison question test for deception," *Legal and Criminological Psychology*, 2009, vol. 14, pp. 279–92.

"These can be good guides . . ." (p. 258): The underlying reason pupil dilation is so useful is because it provides us a subtle window into how much effort a person is making: Pauline van der Wel and Henk van Steenbergen, "Pupil dilation as an index of effort in cognitive control tasks: A review," *Psychonomic Bulletin and Review*, 2018, vol. 25, pp. 2005–15.

"For instance, studies show that children . . ." (p. 260): Kristin Buss, Richard Davidson, Ned Kalin, and Hill Goldsmith, "Context-specific freezing and associated physiological reactivity as a dysregulated fear response," *Developmental Psychology*, 2004, vol. 40, pp. 583–94.

"We see a similar inappropriate response . . ." (p. 260): Jonathan Rottenberg et al., "Sadness and amusement reactivity differentially predict concurrent and prospective functioning in major depression disorder," *Emotion*, 2002, vol. 2, pp. 135–46.

"Based on this type of evidence . . ." (p. 261): Jonathan Rottenberg and

Alexandra Hindash, "Emerging evidence for emotion context insensitivity in depression," *Current Opinion in Psychology*, 2015, vol. 4, pp. 1–5. Jonathan Rottenberg has written a compelling and highly readable book on this provocative perspective on depression: *The Depths: The Evolutionary Origins of the Depression Epidemic*, Basic Books, 2014.

"In an interesting study, psychologists gathered three groups . . ." (p. 261): Yair Bar-Haim, Talee Ziv, Dominique Lamy, and Richard Hodes, "Nature and nurture in own-race face processing," *Psychological Science*, 2006, vol. 17, pp. 159–63.

"A fascinating study shows that preference . . ." (p. 262): David Kelly et al., "Three-month-olds, but not newborns, prefer own-race faces," *Developmental Science*, 2005, vol. 8, pp. 1–8.

"In what has been called a 'categorization instinct' . . ." (p. 263): My husband, social psychologist Kevin Dutton, coined this term "categorization instinct" in his book *Black and White Thinking: The Burden of a Binary Brain in a Complex World*, Transworld Publishers, 2020.

"Social psychology journals are weighed down . . ." (p. 263): An early study of "out-group homogeneity effects" is reported in Edward Jones, George Wood, and George Quattrone, "Perceived variability of personal characteristics in in-groups and out-groups: The role of knowledge and evaluation," *Personality and Social Psychology Bulletin*, 1981, vol. 7, pp. 523–28.

"This has been shown in a series of studies . . ." (p. 263): Patricia Linville, "Self-complexity as a cognitive buffer against stress-related illness and depression," *Journal of Personality and Social Psychology*, 1987, vol. 52, pp. 663–76.

"Indeed, some studies have shown . . ." (p. 268): A German team of psychologists showed that positive mood states led to improved intuitive judgments: Annette Bolte, Thomas Goschke, and Julius Kuhl, "Emotion and intuition: Effects of positive and negative mood on implicit judgments of semantic coherence," *Psychological Science*, 2003, vol. 14, pp. 416–21.

"When we are tired and distracted . . ." (p. 268): Several studies show that we are at our most creative and intuitive when we are tired. There is a good overview of this work in an article by American psychologist Cindy May: "The inspiration paradox: Your best creative time is not when you think," *Scientific American*, March 6, 2012.

Conclusion

"It's no surprise, then, that there's lots written on grittiness . . ." (p. 272): In her compelling book, psychologist Angela Duckworth describes her extensive research around the benefits of grit for children and for adults: *Grit*, Scribner, 2016.

"Sprinkle a little awe . . ." (p. 278): There is growing evidence that experiencing awe has positive benefits for our health and happiness. You can find a good overview in an article written by Summer Allen for the *Greater Good* magazine, "Eight reasons why awe makes your life better," September 26, 2018.

"There are many examples of how people who are generalists . . ." (p. 278): You can read about many fascinating examples in the bestselling book by David Epstein, *Range: How Generalists Triumph in a Specialized World*, Riverhead Books, 2019.

ACKNOWLEDGMENTS

The kernel for *Switch Craft* formed many years ago and has taken a long and winding path to get to this stage. It has truly taken switch craft, and boundless agility, to write a book that I hope people will enjoy and find useful. I am grateful to so many who have helped along this journey.

First and foremost, my wonderful literary agent and friend, Patrick Walsh. Patrick always believed in the project and managed to wrestle a long and unwieldy set of ideas into something that made more sense. Thanks, Patrick, for the constant support and friendship over the years. I am also indebted to Kirty Topiwala, my editor at Hodder whose enthusiasm for *Switch Craft* motivated me to push further. Kirty championed *Switch Craft* from the beginning and her gentle encouragement to scale back here and expand a little there transformed the manuscript for the better. It's been a pleasure working together and I hope we get to do it again someday. Kirty had to drop the reins as the manuscript was nearing the end to go and help populate the world. Those reins were brilliantly taken up by Anna Baty, whose careful reading and incisive editorial comments helped us reach the finish line. I am grateful to the entire team at Hodder, whose enthusiasm for *Switch Craft* kept me going through many an early morning writing session. I am also hugely thankful to Gideon

Weil, my editor at HarperOne in the US, whose early comments and unbridled belief in the power of switch craft helped to shape the book.

My enduring thanks go to the many collaborators past and present, whose ideas and discussions have helped to shape my thinking and open my mind about emotion, feelings, and how they impact on how we think. These include Lisa Feldman Barrett and her team in Boston, as well as Naz Derakshan and her team in London, along with many from the "cognition and emotion" community including: Yair Bar-Haim, Eni Becker, Simon Blackwell, Andy Calder, Patrick Clark, Tim Dalgleish, Rudi DeRaedt, Chris Eccleston, Ben Grafton, James Gross, Colette Hirsh, Emily Holmes, Jennifer Hudson, Ernst Koster, Jennifer Lau, Colin MacLeod, Andrew Mathews, Lies Notebaert, Hadas Okon-Singer, Mike Rinck, Elske Salemink, Louise Sharpe, Reinout Wiers, Mark Williams, Marcella Woud, Jenny Yiend.

I would also like to thank many of those who have been an essential part of my own lab group, the OCEAN lab, in Oxford: Emilia Boehm, Charlotte Booth, Luis Casedas Alcaide, Rachel Cross, Keith Dear, Hannah DeJong, Alessio Goglio, Maud Grol, Sam Hall-McMaster, Lauren Heathcote, Matthew Hotton, Rob Keers, Anne-Wil Kruijt, Michele Lim, Danna Oomen, Sam Parsons, Anne Schwenzfeir, Annabel Songco, Olivia Spiegler, Desiree Spronk, Laura Steenbergen, Johannes Stricker, Eda Tipura, Ana Todorovic, John Vincent, Janna Vrijsen. Special mention goes to Alex Temple-McCune, who sadly passed away at the age of just twenty-six as I was finishing *Switch Craft*. The way in which Alex dealt with his illness and a constant stream of bad news was a true inspiration to all of us.

This book would not have been possible without the hundreds of participants that have taken part in my various research studies over the years, and I am grateful to all of them for their contributions for very little personal reward. The stories and experiences of many sports, business, and military people whom I have worked with over the years populate these pages and I am thankful to all of them for their honesty and openness in helping us find ways together to enhance performance. Many friends from the world of sport have also helped to sharpen my thinking by asking probing questions and giving lots of examples from their own experiences of why agility matters. These include Joey Barton, John Collins, Sean Dyche, Eddy Jennings, Ronnie O'Sullivan, Iwan Thomas, Harvey Thorneycroft and his "Brilliant Minds" team, as well as Jon Bigg and his fantastic group of athletes down in Sussex including Elliot Giles, Charlie Grice, and Kyle Langford.

Finally, enduring thanks and appreciation go to the person I love most in the world: my husband Kevin. His ability to weather the many storms that have hit over the last few years constantly amazes me and he is my guide, companion, and inspiration in everything I do. He read many sections, suggested the title, offered numerous bits of advice, did some savage editing, suggested stories and anecdotes, made me tea, and generally kept me sane.

INDEX

Page references for notes are followed by n

ABOUT THE AUTHOR

Elaine Fox, PhD, is a psychologist, author, and the head of the School of Psychology at the University of Adelaide, Australia. Prior to her move to Australia, Dr. Fox founded and directed the Oxford Centre for Emotions & Affective Neuroscience (OCEAN) at the University of Oxford, a renowned research center exploring the nature of resilience and mental well-being. A cognitive psychologist by training, she is a leading mental health researcher combining genetics, psychology, and neuroscience in her work. Dr. Fox also runs Oxford Elite Performance, a consulting group bringing cutting-edge science and psychology to those at the top levels of sport, business, and the military. Her 2012 book *Rainy Brain, Sunny Brain* is an international bestseller.